MINDFUL SCEPTICS
GUIDE TO...

food, people and soil

Mark Dangerfield
Christopher M Scott

Mindful Sceptics

Published by Mindful Sceptics

Mindful Sceptics

Springwood, NSW 2777, Australia

mindfulsceptics.info

AI Use Statement

This book was conceived, written and edited by a human who occasionally used AI tools to assist with grammar, editing, summaries and focused searches.

Contents

Preface to the Series V

Preface to this Guide VII

1. Where do plants grow best? 1
2. The intimate connection between soil and global challenges. 5
3. Turning soil into agriculture factories 17
4. What has happened to food prices over time? 25
5. Four scary graphs 38
6. Soil degradation is frightening 52
7. Can agroecology fix soil degradation? 69
8. Always tackle the social issue first 88
9. An economist, a demographer and a soil scientist walk into a bar 102

Evidence Sources 106

About the Authors 108

More Mindful Sceptic Guides 110

Mindful Sceptics 111

Being a Mindful Sceptic 112

Preface to the Series

Ever feel like you're drowning in information and controversy, unsure what to believe? You're not alone. In this age of information overload, finding clarity can seem like searching for a needle in a digital haystack.

But what if I told you that the key to navigating this chaos lies within you?

Welcome to the Mindful Sceptic Guides, your compass in the turbulent waters of modern discourse. These aren't your average self-help books or dry academic tomes. They're intellectual adventures designed to awaken your innate curiosity and sharpen your critical thinking skills.

Imagine being able to cut through the noise, to see past the spin and the hype. Picture yourself in conversations where your insights turn heads, where your words carry the weight of well-reasoned thought. That's the power of being a mindful sceptic, a power that's within your reach.

In the pages of *The Mindful Sceptic Guides*, you'll find a delightfully eclectic mix of crucial facts, thought-provoking ideas, and paradigm-shifting perspectives. We'll tackle big issues with a unique blend of scientific rigour and philosophical inquiry. From conservation to social justice, from food security to healthy diet, economic theory to personal growth, no topic is off-limits.

But here's the kicker. This isn't just about personal enlightenment. The stakes are higher than that. In a world grappling with complex, existential challenges, humanity's survival may depend on cultivating more mindful sceptics—people who can question dogma, seek evidence, and imagine new solutions to age-old problems.

Could you be one of these crucial thinkers? Could your ideas be the spark that ignites positive change?

There's only one way to find out. Dive into these guides with an open mind and a willingness to challenge your assumptions. Prepare for moments of revelation, ideas that will keep you up at night, and insights that will transform how you see the world and your place in it.

Are you ready to embark on this intellectual journey? To become a beacon of reason in a world that desperately needs it? Then, turn the page. Your adventure in mindful scepticism begins now.

Welcome to the Mindful Sceptic Guides. Let's change the world, one critical thought at a time.

Preface to this Guide

Every day, billions of people around the world make choices about what to eat. Whether selecting items from a supermarket shelf or buying food from a street vendor, these individual decisions are part of a global phenomenon that has reshaped our world over the past two centuries.

In *The Mindful Sceptic Guide to Food, People, and Soil,* we explore the critical connection between three elements rarely mentioned in the same sentence: our food, the growing human population, and the soil that sustains us all.

In just 200 years, we've gone from one billion hungry humans to eight billion. That's not just population growth; it's a population explosion fueled by an unprecedented infusion of energy into our food systems. All this food and these people are connected by one thing most of us never think about—soil, the unsung hero of human civilisation. And spoiler alert, the soil is in trouble.

In these pages, we will connect dots you never knew existed. We'll travel from bustling supermarkets to quiet fields, from teeming cities to depleted soils. We'll explore the intricate dance between food prices, population growth, and the earth beneath our feet.

Why? Because these three themes—food, people, and soil—are rarely mentioned in the same breath. And that is not just an academic quibble. It's a potentially catastrophic oversight.

But it is not too late. By understanding these interconnections, we can begin to address the challenges we face in feeding a growing population while preserving the health of our planet. This book is an invitation to open your eyes, to see the world in a new way, and to be part of the solution.

So, are you ready to become a mindful sceptic? To peer beneath the surface of your daily bread into the complex, fascinating, and absolutely crucial world of food, people, and soil? Then, turn the page. Your next meal and the future of humanity awaits.

Chapter One

Where do plants grow best?

Warm, wet rainforest teaming with life is where plants grow best, right? Sure, that makes sense. Only most of the planet is far from lush.

You are a mammal which makes you an animal that by definition, cannot make its own energy to defy entropy. Along with all other animals, you have to eat.

Evolution has resulted in millions of ways to find, capture and consume food with sufficient energy and nutrients to keep animals alive long enough to acquire the next meal. And yet all animals, deep ocean sulphurous vent species notwithstanding, have this one thing in common. It's the same for the hyena, the hippopotamus, and the honey bee—all animal food begins with a plant fixing energy from the sun through photosynthesis.

Terrestrial animals can't get by without plants.

As omnivores able to eat plants and the animals that eat them, humans know all about plants as food and as fodder. So, I will begin this exploration into the odd combination of food, people and soil by asking about plants and where they grow best.

Warm or preferably wet and warm is where plants grow well. Lush, green rainforest teaming with life milling about among buttressed trees so tall their leaves are out of sight. Next, we might imagine a woodland with fulsome trees and carpets of spring flowers or a meadow all green grass and morning dew. Either way, we recognise plants growing well when they are green and luxurious. How little of the earth's land surface supports such profusion is often overlooked.

Here is a map from research that modelled a standard measure of plant growth, net primary production, from a combination of annual temperature and rainfall. NPP is the difference between the energy fixed by autotrophs and their respiration. The map shows the lush, most productive areas of the earth for plants in blue and green, with orange regions the least productive.

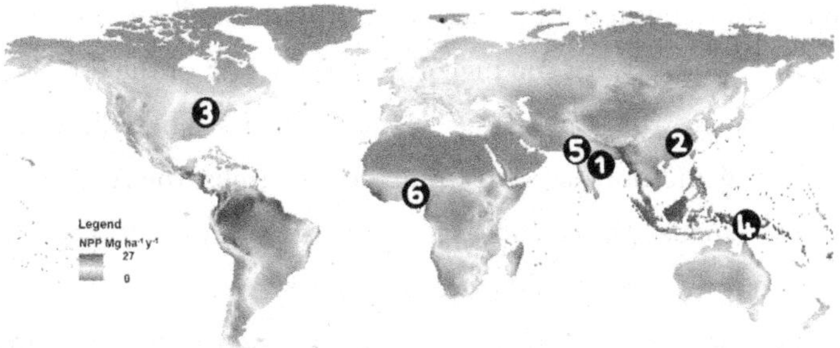

Map of mean Net Primary Production (NPP) in mg C ha−1y−1derived from the mean annual temperature and mean annual precipitation using the Miami model. Source: Abdalla, M., Hastings, A., Chadwick, D. R., Jones, D. L., Evans, C. D., Jones, M. B., ... & Smith, P. (2018). Critical review of the impacts of grazing intensity on soil organic carbon storage and other soil quality indicators in extensively managed grasslands. Agriculture, Ecosystems & Environment, 253, 62-81.

I added numbers on the map to locate the six most populous countries on earth - India, China, US, Indonesia, Pakistan, and Nigeria - that account for 49% of all people alive.

African savanna dotted with a million wildebeests munching away at the never-ending supply of grass is another caricature of plant growth. The Serengeti wildebeest population is a consequence of a unique combination of seasonal rain that forces migration that rests the grass from continuous grazing and the periodic eruptions of volcanoes that replenish nutrients. These open woodland and grasslands comprise much larger areas than tropical forests, the map's light green to yellow areas. There are 53 million km^2 of grasslands, which is equivalent to 40% of the global land area without permanent ice cover.

What is striking about the map is the proportion of the land area that is too dry or too cold or too high or a combination of the three to have much plant growth—the orange to brown shades. The Sahara, Middle East, central and northern Asia, much of Australia, the Andes and the Rockies, Canada, Alaska, and Greenland have little primary production. Most of it is restricted to short seasonal pulses.

At least half of the Earth's land surface is not great for plant growth, nor is it hospitable to humans. The proportion of naturally productive land that can grow food or support vegetation for abundant wildlife is finite. And it's no surprise that most people live where the plants grow best. And because most of us live in or near lush areas, this is our vision of the world and helps us assume that plants will thrive everywhere. The truth is different. Plants grow best on a small proportion of the global land surface.

Chapter Two

The intimate connection between soil and global challenges.

Feeding everyone well requires soil. This universal truth has been with humans on every step out of Africa and across the planet.

Look down

Next time you venture off a concrete path, look down. Under the plants and leaf litter, beneath your feet, is soil. If it's been raining and all you see is mud that will make a mess of your shoes, look again. You may be underappreciating a resource vital to life on Earth. Rather than just something to exercise the bristles on your doormat, the soil is fundamental to our lives because it has a significant role in producing food. A moment's thought and you remember that your primary school teacher said something similar.

You may be less aware that soil also contributes to clean air and water and is home to many of the planet's critters, the soil biodiversity. Then there is the carbon in the soil that powers all the biology and contributes to soil structure by binding soil particles together whilst helping exchange nutrients between the soil and plant roots. This crucial material that scientists call soil organic carbon or SOC combines dead and live organisms in the soil. And whilst, on average, SOC accounts for less than 5% of the mass of upper soil layers and diminishes with depth, it is essential to plant growth. There is a direct link between the amount of carbon in soils and what humans would describe as that soil's health and productivity.

Although soil carbon tends to vary in time and space, adding to the complexity and cost of measurement, we know it is easily disturbed. Managing land for food affects how carbon moves from soils to plants, plants to animals, animals to the atmosphere, atmosphere to plants, and back to soils. Here is the trend that occurs almost everywhere humans use soil to grow food—cultivation depletes carbon levels everywhere scientists have looked.

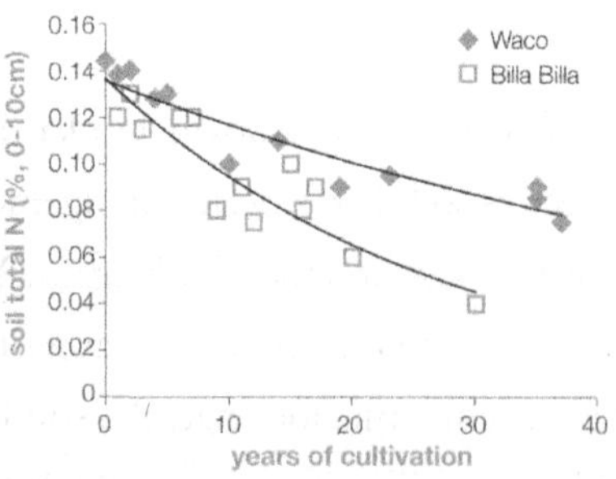

Figure 2: Graph of decline in soil total N with years of cropping. The decline was greater for the Billa Billa soil (clay content 34%) than the Waco soil (clay content 74%).
SOURCE: Based on Dalal and Mayer (1986a,b)

This is not good. It indicates a loss of soil health that is just as critical to the future of humanity as the effects of climate change or biodiversity loss, even the impending decline of fossil fuels. Here is why.

Five essential soil functions

Human ingenuity, technological prowess and a pulse of fossil fuel energy have changed us from a species that lived in nature, fossicking for food and finding shelter where we could, to one that lives off nature. We find and exploit natural resources, converting them into energy and materials that give us a very different life from our ancestors. It also resulted in a massive population spike. Through this energy and population pulse, the soil has become one of the resources to exploit—a natural resource to convert into goods for sale.

This would be just another consequence of the human enterprise and nothing to chirp about except for what soil does. Soil is crucial to the human enterprise. We walk on it, grow our food in it, rely on it to hold and filter our freshwater, build on it and above all, ignore it. Soil is essential to human survival and well-being, yet we take it for granted. Part of this is because the soil is quite robust to the forces of nature and even the machinations of human impacts during the 12,000 years since the invention of agriculture. Most soils can take quite a beating before they show any wear and only collapse after heavy exploitation, such as 30 years of cultivation.

Apparent resilience makes humans complacent; only complacency is careless. Take anything for granted, and you are in for a shock when it's gone. And with soil, we know what we will lose.

Soil delivers five essential functions that make it foundational to the human enterprise and a valuable money machine:

1. Nutrient cycling

2. Water retention

3. Biodiversity and habitat

4. Storing, filtering and transforming compounds

5. Physical stability and support

These functions, which we sometimes call services, do not make money on their own, but they make it possible for humans to consistently and predictably grow crops and rear livestock that do.

Whoopy doo. Soil does stuff, and I should care? Well, perhaps not. History tells us that farmers can harness technology and energy to use soil to grow food and fibre without knowing much about these critical soil functions. Instead, they are experts in cultivating crops and livestock under local conditions. They know what the plants need and when to deworm the sheep. And, along with a bucketload of courage, these skills have been enough.

Even with 8 billion people, sufficient food is grown to feed everyone. At least for now, food insecurity is an access and efficiency problem that is fixable with economic, social, and political solutions. It's currently not a production problem. If agricultural production works when the focus is on the crop plant or the livestock with little or no attention to the soil, why should consumers be concerned?

In short, **soil degradation**. But for now, we will even ignore that giant elephant pooping on the carpet. Instead, let's look at the soil and six global challenges where taking good care of soil is all or part of the solution. The global challenges are

- food security
- water security
- climate change abatement
- ecosystem service delivery
- biodiversity protection and
- energy sustainability.

These global challenges are the big ones that cannot be wished away or miraculously solved by the mystical powers of the market. Nor can they be met by sinking more oil wells and making more people. Six of the big ones must be actively addressed, or society will collapse. Let's peek at the intimate connection between soil and these global challenges.

Soil and food security

People are food secure when availability, access, usability (utilisation), and food stability are consistently met for themselves and their families. When you go to the supermarket or your smallholding, there is enough readily available food to meet nutritional requirements whenever you need it.

The vast majority of global food production to deliver food security comes from vegetation grown in soil. Fishing, hydroponics, and aquaponic systems exist and are essential to some. However, soil remains the primary growth medium for almost all human food and several other vital resources, including biofuels, fibre crops and timber. Food quantity, quality and stability require that soil is in good shape and functions consistently even when the climate is off.

Food also needs to be nutritious—including a suite of micronutrients—because food security is not just about the amount of food but also about achieving a healthy and sustainable diet. Food also needs to be safe. There is no use growing giant pumpkins in Alabama if they are contaminated with heavy metals. Even though food security is much more than food production, there is no food security without it. So, food must be grown in quantities that can readily meet demand.

By the way, if you need more on this topic, we have a *Mindful Sceptic Guide to Food Security* covering how people get enough nutritious food.

Soil and water security

Freshwater has always been a critical and often limiting resource for humans. In the hunting and gathering days, people had to live close to springs, streams, rivers or lakes, and when they moved, they needed to know the next source would be within a day's walk. How they found springs in the desert is miraculous, but they did. Local knowledge, memory and cooperation meant that elders or keepers of the knowledge knew where there was drinking water, even in dry times. Often, it was the soil that saved them.

Soil is a sponge. It soaks up rainwater and stores it in the landscape, releasing it slowly to keep streams running and allowing water to seep into the deeper rock layers to pool as groundwater. After the rain falls and percolates through the soil and rock layers, it is filtered, reducing contamination. Water regulation properties of soil that depend on the soil type, the amount of vegetation cover, and the organic matter levels are critical to how much water is retained in the landscape.

Soil and climate change abatement

Unless you live under a rock or attend QAnon meetings, you will know that the issue of human-induced climate change is real and immediate. Acute climate change is with us because the level of greenhouse gases in the atmosphere has increased dramatically over the last 200 years and rapidly in the last 50.

In 1800, the levels of CO_2 in the atmosphere were 283 ppm, and in 2022, it reached 421 ppm, with a current net annual increase in atmospheric CO_2 of around 2.5 ppm. In roughly 200 years, methane concentrations have more than doubled, and nitrous oxide levels have increased by 50%. Agriculture and land use change are responsible for a quarter of these global anthropogenic greenhouse gas emissions.

Whilst the numbers can be confusing and will be endlessly debated, it is enough to realise that greenhouse gases are critical to the climate, have more than doubled since the start of the industrial revolution, and that agricultural practices of land clearing, fertiliser use and rearing ruminant livestock account for at least a quarter of the increase. But here is the thing. Even with these changes, there is more carbon in the soil than in the atmosphere. Soil is a vast reservoir of carbon.

Carbon as dead organic matter and the living roots and microorganisms, collectively known as soil organic carbon (SOC), is by far the largest pool of organic carbon and globally contains over 1,550 Pg C, followed by the soil inorganic carbon (SIC) pool (750–950 Pg C) and terrestrial vegetation (600 Pg C). Therefore, the total soil carbon pool is about four times larger than the terrestrial vegetation and three times larger than the atmospheric carbon pools.

Carbon that plants sequester through the process of photosynthesis is transferred to soil as root exudates, carbon-rich materials that stimulate microbial activity to release the nutrients plants need for growth, the roots themselves that die and decompose to carbon compounds, leaves and branches that fall onto the soil surface and mixed into the soil layers by the action of soil animals. These carbon inputs are an energy source used by bacteria, fungi and a host of soil animals that create complex food webs, each organism using up some carbon in its metabolism and releasing it as CO_2 back into the atmosphere. This essential process we call decomposition.

How soil is managed can mean it is a net sink or source of greenhouse gases. Clearing vegetation and exposing the soil to the elements means carbon is lost. Cover the soil with perennial plants, especially those species with deep root systems and retain some of the biomass that grows and carbon is added (sequestered) into the soil. I explain how modern agriculture has created a soil carbon debt in the *Mindful Sceptic Guide to The Soil Carbon Debt*. Still, for now, it is enough to know that management of the soil for climate change abatement can reduce greenhouse gas emissions or sequester carbon, or both.

Soil and ecosystem service delivery

The goods and services humans need from nature are the consequence of other species. Some we eat, some we use to build our homes or fuel our fires, and some we admire. Then there are all the species that must be present to clean the air and the water to make life possible. These species moderate and deliver critical ecosystem services. Most of them live all or part of their lives in soil—soil biodiversity is vital to human wellbeing.

Established science tells us that soils that store and receive regular organic carbon inputs have a rich diversity of bacteria, fungi, insects and non-insect invertebrates that surpass the rainforests for species richness and complexity. These organisms create complex food chains that decompose organic matter to release nutrients essential for plant growth. Humans cannot live without them.

Imagine for a moment what would happen if decomposition stopped. Which of these outcomes is most likely 1) we would be wading through knee-high piles of leaves and twigs, 2) nothing changes because the rain washes the leaves away, 3) all the plants die, and the animals starve. I think you know the answer.

Soil and Biodiversity protection

In a teaspoon of soil, there is dirt (sand, silt and clay), soil carbon (some dead plant material, dead animals, and fecal matter), and water. Then there are the organisms, the life that turns dirt into soil. On the teaspoon, there are 100+ million bacteria, 150+ micrograms of fungal hyphae, 10,000+ protozoa, 100+ nematodes, 150+ Arthropods (mites, collembola, insects), and 10+ other invertebrates (earthworms, woodlice, centipedes, etc).

At the taxonomic resolution of species, the one we are familiar with for birds, mammals, and most plants, a teaspoon of soil typically contains thousands of species, many not yet described by taxonomists. Then, there is diversity at the genetic level that is so complex that science has yet to get a good handle on its scale, let alone start a realistic catalogue. What we do know is that **all soil is hugely biodiverse**.

Soil is the habitat for the most significant gene pool and diversity of species, which enables plant growth, the recycling of waste and the provision of nutrients affecting food and water security. Soil is a habitat for biodiversity and the foundation for the above-ground habitat, home to the more familiar birds, mammals and plants Sir David Attenborough tells us are biodiversity. All the chimps, zebras, kangaroos, koalas, and pandas exist because their food grows in the soil. Soil is habitat, and it supports habitat.

Soil and energy sustainability

Humanity has gorged on the near-free fossil energy of coal, oil and gas—a one-time gift from Gaia that humans have converted into food and drink and made more humans. Fossil fuel resources are finite and come with externalities (waste and pollution) that the planet can absorb for a time but not indefinitely. This is why humans have, somewhat reluctantly, set an energy transition in progress. Over the coming decades, society will find alternatives to fossil energy.

One option is using plants for energy production, for example, to extract oil from seeds using modern industrial techniques. These bio-fuels are not direct substitutes for oil, not least because they have a chemistry that corrodes pipes and engines much faster than kerosene, gasoline or diesel fuel, but we will need some of them across the energy transition. Growing plants for energy means they are not raised for food but still need land, soil, nutrients and water. The trade-off with food security is acute.

Before the Industrial Revolution, humans used wood and charcoal for heat and cooking. In principle, trees are renewable, and many humans will return to this energy source when they can no longer afford the electricity, gas and coal they use now. People will cut down and burn trees and shrubs to keep themselves warm and cook their food if the transition is too slow or patchy for energy prices to stay within bounds.

A glance at soil and global challenges

Humanity cannot exist without the soil. Even this cursory glance at the soil and global challenges selected from what could have been a very long list demonstrates how central soil is to what is coming and how humans have managed to become so numerous and, for the most part, prosperous. Any solutions we find for food security, water security, climate change abatement, ecosystem service delivery, biodiversity protection and energy sustainability must consider soil.

Managed well, soil can be an ally in meeting global challenges, such as helping to fix the climate with soil carbon sequestration. Ignore it, and the challenges get deeper.

Soil is life. Take soil for granted, and humanity is doomed to collapse. Pay attention to the soil to manage it wisely, and it can sustain the current and future populations.

But just saying these things and listing a few apparent reasons why soil is at the heart of global challenges is not enough. Soil has to become as important to each consumer as their pay packet. They must understand that there is no food, clean water, and pay without healthy soil. The foundation of people's lives is at risk if we let soils degrade.

The reality is that fixing challenges like climate change and biodiversity loss cannot be done without healthy soil. However, fixing them becomes irrelevant without food or ecosystem services. So when people say climate change is **the** critical issue of our time, nod wisely. They are right. It is vital but not **the** critical issue. That unwelcome accolade goes to maintaining healthy soil.

Key Points

- Soil is a vital resource essential for life on Earth, contributing not only to food production but also to clean air and water and serving as a habitat for organisms.
- Even a tiny amount of soil contains diverse organisms, crucial in decomposition and nutrient cycling processes.
- Soil organic carbon, a mix of living and dead organisms, is vital to soil health and productivity.
- Soil is critical in tackling significant global challenges, including food security, water security, climate change mitigation, ecosystem services delivery, biodiversity protection, and energy sustainability. Soil is a foundation for various ecosystem services such as nutrient cycling, water retention, biodiversity support, and physical stability, contributing to human well-being.

In the next chapter

Now that we know the soil is essential, we look at how humans have used soil as a factory to grow food and fibre that feeds and supports over 8 billion people, their livestock, and their pets.

Chapter Three

Turning soil into agriculture factories

Add energy and fertiliser to the soil, and it becomes the agriculture factories that feed half the global population. However, there are problems with industrialisation because soil is fragile.

The factory floor

The birthplace of the iconic Ford Model T automobile, the Ford Piquette Avenue Plant in Detroit, Michigan, still stands today as one of the world's most significant automotive heritage sites. Built in 1904 in the neighbourhood of Milwaukee Junction, the late Victorian-style brick building was modelled after New England textile mills and measures 402 feet long and 56 feet wide over three stories, with 67,000 square feet of floor space. There are 355 windows, which provided light and ventilation in the days before air conditioning.

Moving fast on the success of the Piquette Avenue plant, in September of 1907, Henry Ford purchased a 130-acre tract of land in Highland Park, Michigan. On this site, the Ford Motor Company built the factory that contained Ford's moving assembly line to mass produce the Model T. Assembly-line production allowed the price of the Model T touring-car version to be lowered from $850 in 1908—equivalent to about 18 months' salary for an average wage earner—to less than $300 in 1925 or just four months' salary. And so began the phase of the industrial revolution fuelled by oil.

The Ford buildings in Detroit and Highland Park no longer house machines and workers who make cars. That function has passed. The greater efficiency of technological innovations and scale replaced these modest buildings with bigger and bigger operations. Modern car plants are more extensive, brighter and more productive than Henry Ford could have imagined. They pull parts and materials from a complex supply network and assemble cars with up to 30,000 individual parts at great speed and efficiency until a microchip shortage occurs. What factories did was show how practical the application of energy to resources could be in making goods in vast volumes by precisely controlling the process.

This anecdote aims to say that humans are genius at controlling the environment for their ends—the ultimate expression of command and control. Unsurprisingly, we applied this control to food production. When the Piquette plant was buzzing, many US and European farms were converted from small, nutrient recycling operations run on human and animal power with minimal inputs into agriculture factories run with machines and energy inputs. Not least among these inputs was the Fordson, the first tractor that combined a small size, lightweight and manoeuvrable frame with affordability. The Fordson tractor went into mass production in 1917, just a decade after the Highland Park factory opened, and became the first tractor that was cheaper to maintain than horses.

Today, we take industrial production for granted, but does the analogy of the factory hold for the future of food?

The factory

Capitalism is super successful at converting energy into work. Tasks that are too hard, repetitive or exhausting for people or animals are performed by machines faster and more efficiently. The factory is its expression, the engine room of the process that now includes agricultural factories that apply production efficiency to food.

A factory is a large industrial facility or complex specifically designed and equipped to carry out various stages of production, manufacturing, or assembly processes on a large scale. The primary purpose of a factory is to efficiently transform raw materials, components, or semi-finished goods into finished products through a combination of machinery, equipment, and skilled labour.

Suppose you want to manufacture a motor vehicle that customers can purchase at a realistic price. As Henry Ford so famously figured out, you will need a production line where the vehicle parts are logically bolted, welded, pressed and sprayed onto a chassis by a host of skilled workers. The production line turned a jumbled mass of components into a shiny new model T Ford—resources plus energy and labour go in and out comes a salable product. Modern production lines have honed this basic idea to add exceptional efficiency with the help of computers and robots.

In factories, humans direct systems that convert energy and raw materials into goods. Production happens at scale to make the process as efficient as possible, with profit as the driver of this efficiency. Capital is deployed, human and fossil labour harnessed, and consumers can access goods that make their lives easier. All that consumers have to do is pay for the privilege and then go to work to earn the money to make the payments.

This all works splendidly for consumer goods that have become familiar to everyone as the Industrial Revolution progressed. Cars, obviously, but also everything else in the home, from garbage bags to flat-screen televisions.

Fields as agriculture factories

Fordson tractors might be cheaper to run than dreys. Still, they are quaint compared to what happened after World War II when the Green Revolution in agriculture applied the efficiencies of factories to food production. Along with mechanised food processing and supply chains, farmers had access to extensive irrigation, chemical fertilisers and pest control, crop varieties resistant to diseases and drought, high mechanisation, adaptability to industrial processes, and even gene manipulations.

Overall, it was a technology and science revolution with crop varieties chosen in the lab, mainly based on productivity. It was introduced into fields through mechanised tillage, fertilisers, and pesticides to standardise soil conditions across vast areas. Crop production became an industry.

The livestock sector also intensified with production facilities bringing animals food, water and shelter rather than letting them wander to find sustenance. These high-density facilities are often called factory farms to remind us what they are. Nutrients and energy go in—protein comes out.

Modern technological innovation, especially the advent of satellite navigation and drones, has added layers of precision to these systems, with remote and on-ground sensors feeding information on soil and plant properties into strategies that can optimise the allocation of water, nutrients and pesticides. All over the world, many farms have intensified production and become factories.

Factories offer control

Here is the thing about factories—they are super efficient at converting raw materials and energy into goods. Production efficiency comes from consistent, conforming, and predictable processes to convert raw materials to goods. Any risks that could upset efficiency are easily recognised and quickly neutralised, with breakdowns rapidly located and fixed. Everything is under control.

And agriculture factories are no different. Farmers make fields and farms as uniform as possible to consistently control the growth of plants and animals. They use machines to till the soil to get rid of the stones, and break up the clods whilst getting rid of the existing weeds. The spraying of chemicals dispatches any weeds that arrive on the newly ploughed soil. Nutrients are applied and, if necessary, water through irrigation as the plants grow—the whole system is designed to channel the energy and inputs into usable crops and livestock products. Add economies of scale to this approach, and intensive agriculture can become profitable, especially if the commodity produced is a raw material for processing into cheap food—grains, sugar and seed oils.

Over half of global agriculture is food production subsidised by energy and nutrients. And humanity cannot do without this system. It feeds billions of people, especially urban dwellers, who need help providing food for themselves. Mostly, it feeds them reliably, consistently and with foods they like. These humans have promptly taken the vast amounts of food that the efficiency of industrialised agriculture produces and **converted it into more humans**.

The consequences of agriculture factories

At least half of the 8 billion humans alive are fed by intensive agriculture. These people do not need to spend time in the fields or the food processing plants; they can buy their food from markets, stores, or fast food outlets. Purchasing food frees their time to add energy and intellect to the broader economic activity and trade their time for money so that they can buy food.

All this is a clever virtuous cycle for the capitalist. Keeping this economic engine going justifies the command and control system applied to the agricultural factory to turn financial and natural capital into profit. The inputs of labour, energy and nutrients to agriculture factories and their outputs of food commodities are easily understood and measured. Output goes to consumers, who go on to make more consumers. What's there not to like? But what about the factory itself? What is the agricultural equivalent of the factory floor?

Soil is the factory floor of agriculture. It is where almost all the production happens. Only it is far from the inert shell that keeps the rain and snow from clogging the machines on the car assembly lines. It is alive. Soil lives and breathes and can suffer from ill health, stress and overwork.

Soil is also the equivalent of the factory worker replaced by machines. For a time, the replacement is more efficient, works all hours without fatigue and increases production—until there is a microchip shortage. If the soil is where inputs become outputs, and food production becomes dependent on inputs—fertilisers, pesticides and mechanical force—then we had better be sure that the supply of inputs is consistent. Any shortfall quickly becomes catastrophic.

Soil is not truly analogous to the factory building, tempting as it might be to treat it like one. A better way to view soil is as the workers in the building who can renew and restore themselves organically. Soil that can store organic matter, recycle nutrients and hold on to water can renew itself. Perhaps even sustainably.

But I'm drawing a long bow here. Soil is not an array of agriculture factories, even if intensive agriculture uses soil as though it were, depleting the very capacities that make it possible to grow food. And if the FAO is even close to correct on their estimate that 40% of global soils are degraded, then heaven help us, as my gran used to say.

The production line that Henry Ford perfected turned a jumbled mass of components into a shiny new model T Ford or Fordson tractor—resources plus energy and labour go in and out comes a salable product. Factories gained efficiencies that made Western society possible. We then applied the same ideas to agriculture factories and grew vast quantities of food that became vast numbers of people. It's all good and hopelessly precarious.

Key Points

- Industrial factories, like the Ford Piquette Avenue Plant, used energy and technology to increase production efficiency, and innovations applied to food production turned farms into agriculture factories.

- The Green Revolution after World War II introduced mechanised processing, irrigation, chemical fertilisers, pest control, and genetic manipulation to crop and livestock production.

- Whilst soil might be seen as the "factory floor" of agriculture, it is a living system that can be stressed, and its health is crucial for efficient food production.

- Dependency on consistent inputs of nutrients, energy, and resources to maintain productivity poses a significant challenge to sustainable food production and agricultural efficiency.

In the next chapter

One consequence of agriculture factories for billions of people is that they can go to the store and purchase their weekly groceries. If they have money, they don't need to grow their food. Next, we look at what it costs them and find some surprises in the price of food over time.

Chapter Four

What has happened to food prices over time?

Your weekly grocery bill might seem larger than ever as inflation bites everyone. What has happened to global food prices over time might surprise you.

Food prices

Humans are odd creatures. Take me, for example. I love pickled onions. I always have, even when, as a kid, I picked the cooked onions out of any dish my mother served up. Pickle them, though, and I couldn't resist. My current favourite variety is *Three Three's Sweet Peppered Onions* because they are always crisp and have just the right balance of sweet and tangy. On special in my local supermarket here in Australia, they cost a little under $5 for a 500g jar that lasts me about a week. The cost is absorbed into the much larger weekly grocery bill because it is less than 2 per cent of the total, not material in accounting terms and well within the inter-week variation. Although, tragically, it would be one of the first items to go in a budget squeeze.

According to the Australian Bureau of Statistics Household Expenditure Survey, which tracks food prices over time, the median Australian household spend on food and non-alcoholic beverages in 2016 accounted for 17% of total expenditure, a decline of 3% since the mid-1980s. It was still at 16% in 2021. Over several generations, Australian families have spent around a sixth of their family income on food.

In 2021, the average Australian household spent $108,291 annually, equivalent to $2,083 weekly. The weekly spending on food for the home and from cafes and restaurants was $341. Such numbers would have been unthinkable to the householders of the 1980s scrambling to make the budget meet the 15% interest rates. But, here is the thing. Whilst the dollar spent on food has risen as a gross sum since the 1980s, the percentage spent has declined. This means that relative to overall wealth, what Australian families spend on food has declined.

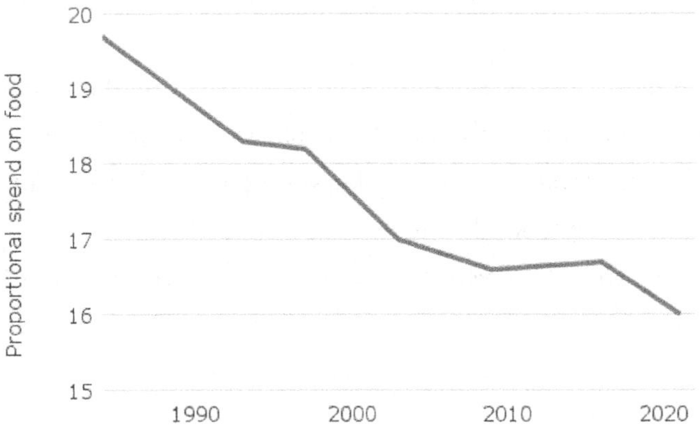

Spend by Australian households on food as groceries, cafes and restaurants as a proportion of available income. Data source: Australian Bureau of Statistics Household Expenditure Surveys

It might feel like food is more expensive, but not so. As Western wealth has grown, the relative cost of food hasn't grown with it. A proportional spend of around 15% relative to income is consistent across developed economies. Around a billion people get their food for a song.

The Global South

Spending less than a sixth of income on food applies to the developed economies. Elsewhere in the world, the situation is different. A typical Nigerian family will spend 40% of their income on food, similar in Kenya, Cameroon and Algeria. In South America, a typical Peruvian family spends 29%, as do families in Venezuela, Columbia and Bolivia. In many Asia countries, the spending is over a third of income, including the two most populous nations, India and China. Try telling that to the average Aussie household complaining about the rising unit cost of a Big Mac.

A graphic from Washington State University, based on 2008 data, shows the majority of the world's people live in countries with transitional or developing economies. These people spend at least a third of their income on food, close to one-half of household income in low-income countries. This picture is one of several showing the global north and south.

Source: Washington State Magazine

If food prices doubled in Pakistan or Indonesia, many people would be under severe financial pressure to meet all their other essential household expenditures. They could become hungry or destitute in a short time. Most of the incidence of child malnutrition due to insufficient nutrient and calorific intake happens in countries in Africa, parts of Asia and South America, where households spend a third or more of their income on food. Not surprisingly, food prices are crucial to people in many of these countries. They pay attention to food prices, and when they rise, unease and unrest often follow.

A loaf of bread story.

In the late 1980s, when I was a fresh-faced academic at the University of Zimbabwe, a loaf of bread in my local supermarket in Harare cost 30 cents. These oblong objects were quite tasty on the morning they were baked, but by the evening, they were a tad firm, and after a day, it was wiser to use them as household bricks. At 30 cents, though, they were an important staple.

Towards the end of my time in that magnificent and troubled country, the price of a loaf went to 50 cents. There was mayhem. Everyone was livid; it was the only thing people wanted to discuss for weeks. They shouted, protested, and ultimately, had to pay 50 cents. In April 2020, 30 years after my time in the country ended, the long-suffering Zimbabweans who, after knowing the Gods were crazy when a loaf cost Z$1 million during the hyperinflation of 2007 and 2008, were freaking out again as bread went to 3.50 RTGS dollars a loaf, up from 1.80 in just a few days. RTGS is the 'Real Time Gross Settlement' dollar—don't ask.

In the first quarter of 2021, Zimbabwe was in crisis again with runaway inflation, spiralling commodity prices, and a food shortage. The cost of bread reached $98 a loaf in March. A loaf of bread has become a luxury in Zimbabwe and many low-income countries that must import wheat.

Volatile bread prices tell us whether food prices are reasonable, terrible, or somewhere in between, depending on where you live and your relative food insecurity. Add the income disparity present in all countries, and there are people for whom food prices are out of reach living next door to people who can readily fill their shopping trolly with nutritious food.

So, what do you imagine has happened to food prices globally over the last 120 years? Let's have a look.

Global Agricultural Price Index

The UN Food and Agriculture Organisation, FAO, track food prices worldwide in their Food Price Index. The graph below shows the global agricultural price index from 1900 to 2010, pegged at 100 for 1977-1979.

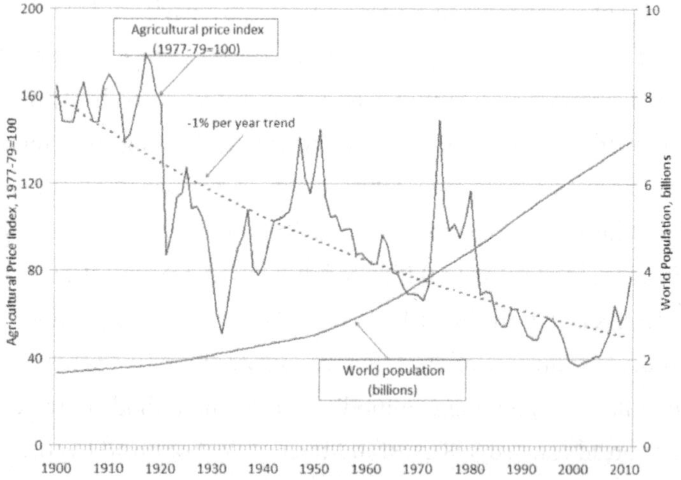

This graph comes from Fuglie, K., and SL. Wang. 2012. "Productivity Growth in Global Agriculture Shifting to Developing Countries". Choices. Quarter 4.

The trend is similar to the proportional spending by Australian families—over time, food has become cheaper. The index is volatile with price spikes, such as the one that followed the oil crisis of the 1970s triggered by the Yom-Kippur War of 1973 and the Iranian Revolution of 1979 that disrupted oil supplies from the Middle East, but then declined by a quarter to the year 2000. The price of food is volatile yearly but **declined** significantly through the 20th century on an exponential trend of around 1% per year. Over the long run and on the average, food became substantially cheaper.

Why the Agricultural Price Index went down

The price index data covers the mechanisation of agriculture, including those Fordson tractors and the Green Revolution of the 1940s and 1950s when irrigation, fertiliser use, new crop varieties, and numerous agronomic innovations boosted crop yields worldwide, especially in Asia. It covers the rise of processed foods from factories, the marketing and retail machinery to sell novel foods to people, and the increase in global container shipping that made food supply chains longer and established the global commodity market. These innovations increased the volume and efficiency of food production that a global market could deliver to consumers at affordable prices. Food became cheaper.

There were price spikes, such as in the First and Second World Wars, the global oil crisis of the 1970s, and the fallout from the GFC in 2008, but the general decline has meant that food shortages and famine have declined. Cheaper food has contributed to well-being and perceived wealth for many people. Steven Pinker suggests that consistent food has helped with the overall decline in global conflicts that, until recently, was a feature of the post-Cold War era. It is also no coincidence that as food prices have dropped, the number of people has risen. Hungry people are angry—well-fed people are much more loving.

What the global population eats is cheaper than it used to be. There is more of it, and the supply is secure enough to keep most people alive. How do we know? Population increases at a rate of 8,000 an hour.

A broad trend always masks any number of nuances. Food has terrible inequalities, and just because calories are grown doesn't make the food nutritious. Still, the truth is that humans are remarkable at growing, processing and selling food—and making more humans.

The reality of food prices over time

Population growth at 8,000 people per hour is a stark and unavoidable reality explained in *The Mindful Sceptic Guide to Escaping the Malthusian Trap*. More and more people, almost all desiring more for themselves, is inevitable. It's not good or bad or moral or otherwise. It is just biology. Unless geopolitical instability delivers an economic collapse or global conflict or the food runs out, this population increase will continue for many decades. Conventional wisdom from demographics has the peak population size at 9 to 11 billion souls sometime in the middle of the century.

Cheap food is essential to keep these people happy, stable and sedentary. If food is expensive or there are price spikes, many will be very unhappy, their lifestyles will be under pressure, and many will be forced to move to find food. The prospects are not good.

In 2020, the FAO revised the base period for calculating the Food Price Index to 2014-16 and expanded its price coverage. Here is the revised graph for the food price index through to 2022. Food was relatively cheaper in the late 1900s, consistent with the previous graph, but in the 2000s, the relative prices began to rise.

Changing the goalposts confuses everything. Pegged at 2014-16 prices rather than 1977-79 is an apples and oranges comparison, and that is why I didn't put the numbers on the same graph. This one tells us that even with the change, food was cheaper towards the end of the 1900s than it is currently and that the volatility is still present. If an index can change by double-digit percentages from year to year, then food prices in the stores and from the vendor carts are changing dramatically, too.

At first glance and proportionately, food prices are rising in trend and through spikes for most people worldwide. Given the acute food security challenges in low-income countries and the increasing inequality in food access in mature economies, such volatility is significant, perhaps crucial to human futures. A brief perusal of human history tells us that food price instability is a huge problem, even the root cause of severe conflicts. Remember that territory is usually the reason states go to war because the land is where the food comes from or where there is the fuel needed to produce it.

Then there is the basic economics. Cheap food is only possible if the supply can meet demand—one of the few fundamentals of economics that is true in our capitalist systems. The price rises if there is an insufficient supply of goods in the right place at the right time to meet demand.

Meeting global demand for food will require ongoing maintenance and gains in the efficiency of food systems across

- Food production
- Harvest and storage
- Food processing
- Food distribution
- Consumption
- Diet and nutrition choices

And here I am thinking that a 50-cent loaf of bread was a problem.

Economists will quickly point out that the history of food prices is on our side. Although demand growth has been with us for a while, for the most part, supply has kept up—nothing like an economic opportunity to drive production volume and efficiency. What most mainstream economists fail to admit is how this miracle happened. How could food production grow and become more efficient whilst meeting the demand for food from 8 billion people—22 trillion kilocalories daily?

The short answer is not industrial intensification or the soil factories or the miracle of the market. It is introductory thermodynamics—**agriculture went from an energy source to an energy sink.**

The food we eat comes from the fossil energy pulse. Roughly half the nitrogen in your body originated in a factory producing fertilisers via the Haber-Bosch process using energy from natural gas. On the farms, all the tractors and harvesters, threshers and quad bikes are powered by oil, as is the fleet of vehicles that move the production off the farm to the market. We grew the food by energy subsidies to the soil and the food supply chain. And now we cannot stop.

The reality humanity faces is to increase food supply and accessibility to cover the locked-in population growth and to maintain these unprecedented levels of production for decades to come—unless the growth is not locked in.

When you go to the supermarket and see the prices up again, and the food budget strained, it is hard to see the general decline in food prices over time. Most of us are blind to the fossil fuel inputs, production innovations, and global supply chains reaching every potential market to deliver cheap food. Complacency is especially true for the lucky ones in developed economies where milk comes from the fridge.

The mindful sceptic might ask if food prices will continue to decline. The answer is no, according to the adjusted FAO index, and even without fiddling with the accounting, ongoing decline seems unlikely. The intensification of agriculture has dramatically reduced yield gaps and generated vast volumes of food transported worldwide. Whether they make money or not, all the innovation options to keep this going will be necessary. The problem comes when the cost of inputs rises, which they will as fossil energy supply wanes. Agriculture can't be an energy sink without a reliable source of usable clean energy and efficient waste management. Neither are part of the current food systems.

It is more likely that the strain from the energy demand of intensification will catch up with production. The absolute amount and the proportional spending on food will rise as the fossil fuel pulse wanes and food production becomes more unstable from climate and geopolitical drivers. Food price rises will be severe in some places and hurt the most vulnerable.

The upside is that people are imaginative, innovative and courageous. Throw a challenge at a human, and she will take it on most of the time. There are already solutions in agroecology and circular production systems, and people can bend their minds to find more. What the fortunate can do is become aware and prepare. Invest in innovation and find new energy sources, improve productivity and reduce waste to make every calorie efficient. There is also an opportunity for social innovation, from policy to practical solutions that tackle inequality, food security and diet. After all, and by definition, food is a people issue.

The mindful sceptic takes a breath.

The technology fix got us here—8 billion growing at 8,000 an hour—but that is not sustainable because there are limits on a finite planet. More of the same in food production is essential for global stability, but we can't have more forever. The sceptic sniffs a problem. It's one gigantic problem with a technical-sounding name that few people know about—soil degradation.

Key Points

- Food prices are volatile but have declined over time, with the proportion of income spent on food varying among households, averaging 10-15% in mature economies and up to 50% in many countries in Africa, South America, and Asia.

- Various factors influence food prices, including population growth, supply and demand dynamics, production efficiency, and global market integration on top of an overall trend from the fossil energy subsidy.

- Affordable food is crucial for global stability, especially as the population continues to grow, and that will need an improved food system from production to consumption.

- Once agriculture shifted from being an energy source to an energy sink, food production increased. Still, there are challenges from this intensification, including rising costs of inputs, energy demands, waste management, climate change, and geopolitical drivers.

- Food prices will rise, especially in regions where many people are already vulnerable.

In the next chapter

Food is about people and all the social constructs that we have invented. It is tempting to follow that truth and uncover details of supply chains, food access, inequity, food waste and even the nutritional quality of specific diets. This would be a logical progression from food prices and people. But in the next chapter, we see why we must return to the basics of where food comes from, back to the soil.

Chapter Five

Four scary graphs

Despite plenty of food, a global surge in hunger, malnutrition, and population trends all circle back to soil degradation.

There is plenty of food.

In mature economies, supermarkets are full of a glittering array of foods irrespective of the season, and access is more a matter of social disadvantage than supply. Indeed, the global north appears to have a problem with too much food, given there is an obesity epidemic. One in three Americans, one in four Brits and at least 10% of just about every citizen in mature economies have a body mass index (BMI) over 30, the metric and threshold the World Health Organisation uses to define the abnormal or excessive fat accumulation that presents a health risk. The WHO estimates that more people are obese than underweight in every region except sub-Saharan Africa and Asia.

So here is scary graph #1

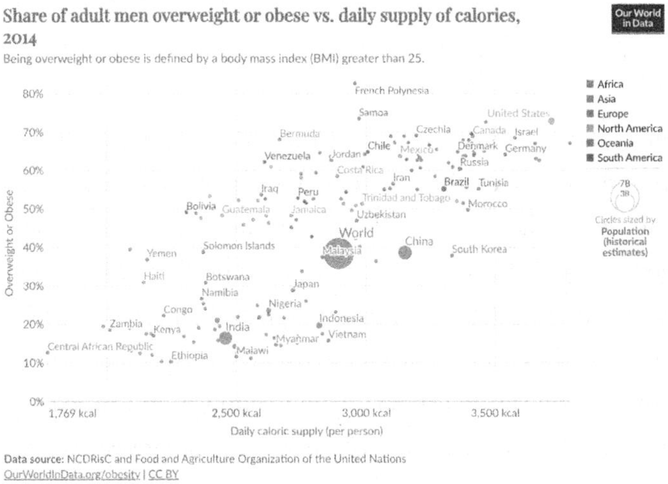

An increasing and significant proportion of adults in mature economies are overweight or obese. The implication is that in many regions of the world, the six-continent global food supply chain delivers more than enough food or perhaps the wrong food. Many people are malnourished because they have too much to eat.

The food supply has also trickled down, at least somewhat. Although famine killed nearly 75 million people in the 20th century, it virtually disappeared in recent decades. The Famine Early Warning Systems Network, created by USAID in 1985, is a leading provider of early warning and analysis on food insecurity, has, in 2023, parts of Ethiopia, South Sudan, Yemen and Nigeria on emergency level of acute food insecurity, but no areas in famine. Fewer food disasters on the news reinforce the perception of plenty of food.

But then there is this headline.

Hunger is rising

Famine might be down, but hunger levels are alarming, with 1 in 10 people on Earth, an estimated 768 million global citizens, undernourished in 2020. Then the COVID-19 pandemic hit. United Nations data suggest hunger increased by 118 million people from 2019, when 8.4% of the world's population was undernourished. Impacts of the COVID pandemic on hunger included food supply shortages, malnourishment due to supply chain disruption, labour shortages due to illness, and a global rise in food prices. These effects will persist for years and have the most significant impact on people experiencing poverty.

Here is scary graph #2.

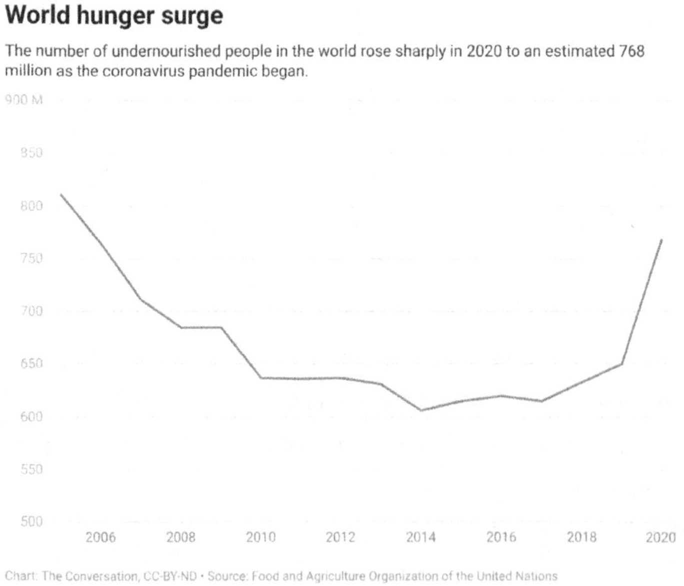

Graph showing the number of undernourished people in the world from 2000 to 2020 with an alarming spike since 2019

As a consequence of our evolution as wandering foragers, we are adapted to persist for a long time, even when our basic level of nutrition is unmet. Humans are an exceptional species that can survive on modest food sources. Such tenacity and physiological resilience. masks the risk. Adding twice the UK population to the hungry in a year means many people were on the edge of hunger. What looked like progress in reducing the number of undernourished people worldwide is reversed in a single disruption. At the sharp end of malnutrition, the World Food Program estimates that in 2023, more than 345 million people faced high levels of food insecurity, more than double the number in 2020. The speed of this rise in undernourished people is alarming, given that hunger is the expression of acute food shortages. The uptick in the scary graph #2 continues.

Then there is the demand from ever more people.

Watch the trends in the population

Genetic and anthropological evidence suggests that *Homo sapiens* have been a species for 300,000 years after separating from our Hominid ancestors. We became a global species during this time and colonised many habitats, but we were never abundant in one place or collectively. Before the invention of agriculture 12,000 years ago, the total number of humans that had ever lived was 9 billion. That is the sum of all the humans who walked the earth for 288,000 years. Since the invention of agriculture, 92 billion people have lived and died. And in 2023, the human population stands at 8 billion or **7% of the humans that have ever been**.

Take your hat off to us. Not only have modern humans persisted, but we are also genius at making more of everything, including more people.

And this ability has accelerated dramatically since the 1950s. When I was born in 1961, 3 billion people were going about their business of making more. In my lifetime, 4.5% of the humans who have lived were born and survived to 2023. This is truly amazing and also scary. Since the 1980s, the absolute number of people added to the global population each year has been around 80 million souls—each year we add the population equivalent of one German or three Australias to the worldwide population. Rather than present the typical hockey stick graph of human population growth that shows our numbers trickling along until the Industrial Revolution sent them skyward, here is another way to visualise population growth.

It is scary graph #3.

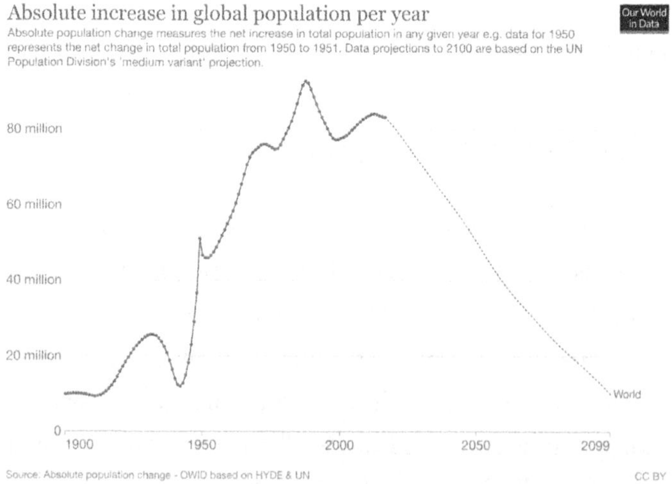

Absolute increase in the number of people on earth each year. It is currently over 80 million even as soil degradation expands

Take some time to peruse this information. Adding 800 million new people in a decade is a substantial extra demand on food supply on the existing daily 22 trillion kilocalories challenge for energy needs. Then, there is nutrition to maintain metabolic health. As the saying goes, man cannot live on bread alone. Agriculture altered the game. We could decouple our numbers from the constraints of a foraging animal and increase our numbers on the energy excess from crops we grew in fields and livestock we herded across grasslands. Then everything changed. Fossils arrived and turned agriculture from an energy source to an energy sink. We used the power of coal, oil, and gas to drive machinery on farms and make fertilisers, herbicides, and pesticides in factories to intensify agriculture—we could and did grow more food than we could eat.

This brings us to scary graph #4.

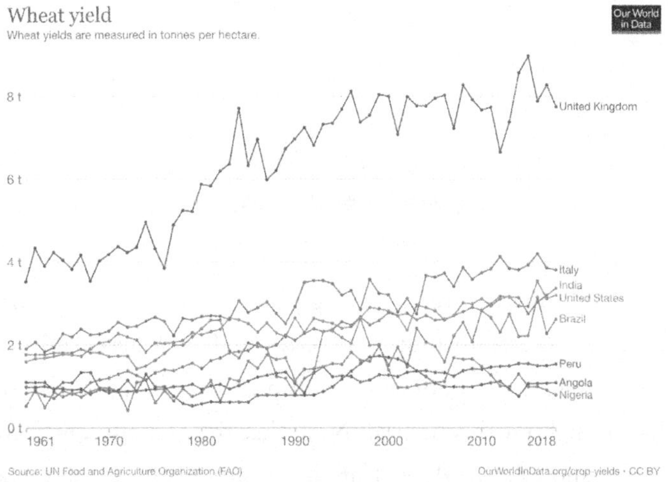

Graph of wheat yields by country from 1960 to 2018

At first glance, this doesn't look scary at all. Graphs showing a trend from bottom left to top right are good and appear in board rooms all the time. In this one, aggregate wheat yield is increasing, reflecting a global trend since the 1960s for greater production and efficiency across most agricultural commodities. But take a closer look at the last 30 years. Here is the graph that zooms in on that period.

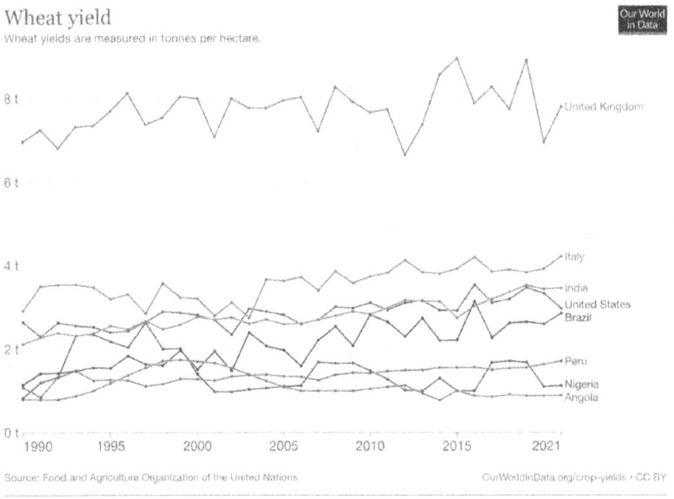

Graph of wheat yields by country from 1990 to 2018 showing that growth in yield has stalled.

Some countries are still achieving yield gains, but most yields have plateaued in several, including the UK, the US, Nigeria and Angola. However, the human population continues to grow at 80 million annually throughout this period. Note that this is a random subset of countries, and Our World in Data has an interactive graph for wheat yield where you can select countries to see the trend from 1961. There is a serious implication if the agricultural revolution has stalled and crop yields have stabilised or reached close to the maximum under the existing technologies for soil management. Yet, there are more mouths to feed—meeting demand will require more land.

No problem. We can always cut down some trees and expand global cropland and pasture/rangeland to add agricultural land to the production system. We already use half the habitable land area, some 49 million km^2 for agriculture, including 11 million km^2 for crops, and all the easily accessible and productive areas. Little land is left unless we cut down forests or expand into shrubland. We already know that human land use is the reason for biodiversity loss and that if we keep cutting down forests to grow food, we increase greenhouse gas emissions and make it impossible to tackle both problems with rewilding. There is the option of rehabilitation of abandoned agricultural land, especially in Europe and the United States, land no longer productive due to, for example, climate-induced vegetation shifts in Siberia and woody encroachment of rangelands in the United States and Australia. Only this is expensive and technically challenging. Realistically, we are in a bind over land use.

A fifth of global soil is degraded

Soil degradation is where the soil is depleted in organic matter, nutrients, and structure to levels that reduce plant growth (biomass production) under typical climate and management conditions—a fancy way of saying that depleted soils grow less food. When the human population is the highest it has ever been—more than three times the number of people to feed than were alive in 1950—and at least half of these people can afford to access nutrient-rich food, the soil that grows the food is failing.

At least 20% of global soil is degraded—one in every five fields.

As we saw, good soil operates as a medium for plant growth. It exchanges nutrients, holds water, is stable over time and maintains a vast community of mostly microscopic organisms, the soil biodiversity, that facilitate its function. Not all soils are equally good. They vary from place to place in what they can offer plants, even within a field. Some are dry and relatively inert, some are perpetually moist and teeming with life and nutrients, and every soil changes with the climate and time. Plants and the animals that depend on them match this variation with their diversity. Matching means that most soils have at least some plants growing in them. Even the driest desert soils have some tenacious grasses and shrubs in favourable pockets.

What we call 'good' is more than this matching of nature to the conditions. Good soil is the productive kind, soil where the plants grow well. Not just any plants but the plant species and varieties that humans find helpful, from melons and maize to marijuana. Degraded soils produce fewer crops and livestock than their potential; they yield less. At the extremes, deterioration means lower productivity and the loss of topsoil altogether from erosion by wind and water. Various scientifically robust estimates have soil erosion rates an order of magnitude higher than that of natural erosion or soil formation processes. Researchers have described soil degradation as a significant environmental issue occurring in various forms. The common ones are:

- Soil erosion happens when wind or water removes the top layer of fertile soil, exposing less fertile layers beneath. It can result from land management practices, deforestation, and construction activities.

- Chemical degradation from excessive use of chemical fertilisers, pesticides, and other agrochemicals alter soil properties, affect soil structure, and decrease natural fertility over time.

- Compaction occurs when the soil particles are compressed, reducing pore space and limiting water infiltration and root growth. Land management practices involving heavy machinery, intensive grazing, and over-tilling can contribute to soil compaction.

- Salinization occurs when the salt concentration in the soil increases to levels harmful to plants. This can happen in arid regions where irrigation leads to the accumulation of salts in the soil.
- Acidification occurs when the pH of the soil decreases, making it more acidic, typically from excessive use of nitrogen-based fertilizers and the presence of pollutants.
- Nutrient depletion can occur with continuous monoculture farming and improper nutrient management.
- Loss of soil organic matter reduces the ability of soil to retain water, provide nutrients, and support beneficial microbial communities.

The scary statistic of soil degradation in a fifth of global soils means that one in every 5 acres of land used for agriculture across the globe is not performing as well as it could because the soil is depleted from overuse or poor management. In some places, this soil degradation problem is already acute enough for farmers to leave the land. Many fields are abandoned each year because the soil is exhausted of nutrients.

> *The evidence suggests agricultural land abandonment is widespread globally, driven by multiple factors, and has strong implications for biodiversity, ecosystem services and human wellbeing*

Prishchepov, A. V., Schierhorn, F., & Löw, F. (2021). .
Land 2021, 10, 97.

I'm guessing that apart from the 'soil is where plants grow' part, you didn't know much about this. Most Western consumers are blissfully unaware of soil because what they see at the store is plenty of accessible food.

But there are some essential takeaways from the four scary graphs:

- At least half the people in the world are malnourished because there is not enough food for some and too much of the wrong food for others.
- The number of people in both categories can rise sharply.
- Despite the challenges of malnutrition, the global population continues to grow at over 80 million people per year—152 per minute.
- In many places, grain yields have plateaued, and soil degradation is starting to impact the food supply.
- These trends do not go together well.

It's a thrill ride. So what, the supermarkets are full. Adrenaline seekers love high-stakes games of risk and reward. There is opportunity everywhere in a world on the brink of regional and even global famines. Heck, when food prices rise, some people will make a dollar. But this neoliberal mindset only applies to the billion or so people on Earth who can afford to take a holiday overseas. The last thing these folks know about is soil degradation, which raised the stakes.

As the late and truly great Professor Hans Rosling so entertainingly explained in 2010, two billion 'global poor' people aspire to a pair of shoes and, maybe one day, a bicycle. And we are locked into that number growing to 4 billion by 2050 unless the miracle of economic growth throws us off a geopolitical cliff. Hans Rosling gives his epic TED talk that explains the what and why of global population growth from 1960 to 2050.

The scary statistic of soil degradation in a fifth of global soils means that one in every five acres of land used for agriculture across the globe is not performing as well as it could because the soil is depleted from overuse or poor management. In places, it is the people experiencing poverty who suffer from this global brinkmanship. They cannot afford for the prices of food to rise or for the supplies of cheap food to run low. The supermarkets are full for the lucky few. Luck makes us blind to the source of that food, the medium that sustains us all, rich or poor. Humanity will need every acre of fertile soil to stay fertile, or those billions will starve. And before they do, in their desperation, they will raid the supermarkets. Think about it.

Key Points

- Food availability in mature economies where supermarkets are stocked year-round is precarious despite the abundance that has given the global north an obesity epidemic, pointing to a problem of overconsumption rather than actual scarcity.

- The coexistence of hunger and obesity is a paradox. While famine has largely disappeared, hunger levels are still alarmingly high, with nearly a billion undernourished. The COVID-19 pandemic further exacerbated hunger, leading to increased food supply shortages, malnourishment due to supply chain disruptions, and rising food prices.

- The global population continues to proliferate with associated food demand that pressures agricultural systems to produce more.

- Soil degradation is a crucial concern, with at least 20% of global soil being degraded and less capable of supporting plant growth and productivity, leading to reduced crop yields and even agricultural land abandonment. Abandoned fields result from soil exhaustion, threatening biodiversity, ecosystem services, and human well-being.

- Sustainable soil management practices to mitigate soil degradation, which include crop rotation, agroforestry, cover cropping, and reduced tillage, are vital to prevent further soil degradation and maintain agricultural productivity.

- Soil degradation is a huge risk to global stability. The vulnerability of food supply chains and the risk of social instability due to food shortages underscore the urgency of addressing soil degradation and ensuring a stable food supply for all.

In the next chapter

We have established that soil degradation is a risk that you should find disturbing. Now we need more detail.

Chapter Six

Soil degradation is frightening

Sometimes it is essential to take stock, to sit under a shady. tree and contemplate what brought you to this tree. Then, decide if the tree will die because of soil degradation. I kid you not.

Knowing yourself

Soil degradation is not mentioned by the thousands of authors who gave us shelves full of self-help books that taunt us toward awareness on every page. Nor is it on the minds of the Insta influencers or the TicToc gurus, all filming advice on how to smell the roses, hear the birds singing, and breathe. Apologies. I jest, at least a little. Mindfulness is not just good; it is excellent. There is nothing wrong with sitting under a shady tree, listening to the birds and the whisper of the wind; nothing at all. Knowing yourself has always been the key to a healthy, fulfilled existence. Indeed, awareness of self is the essential step to understanding everything else. Because there is everything else, contemplation time should include musing on the consequences of what humanity has done and will do in the coming decades because we all live in and of this world. And the world has changed.

Soil degradation frights

I know that learning is supposed to be positive and uplifting, but sometimes the truth is inconvenient and a real Debbie Downer. Rather than sugarcoat, a mindful sceptic accepts pragmatic statements of the facts because, as the motto of the University of Botswana says in the Setswana language *Thuto Ke Thebe* or English, 'education is a shield'.

A mindful sceptic knows that knowledge brings power. So here we go with some terrifying details on the soil. First, let's extend the FAO estimate that **one-fifth of global soil is degraded** with a series of observational and experimental facts collated from the scientific literature by researchers from the UK, Brazil and Poland.

- In the last 150 years, more than half of all soils have been damaged.

- Degradation of soil has been accompanied by the loss of more than 50% of the soil organic carbon (SOC) stock in some cultivated soils, with over 2 billion ha affected globally.

- Soils subjected to degradation become a significant emission source of CO_2 to the atmosphere and have liberated an estimated 176 Gt of soil carbon globally—there is 800 Gt C held in the atmosphere.

- Averaged over the last 150 years, the soil carbon loss rate equates to 1.6 ± 0.8 Gt C yr^{-1} compared to total anthropogenic global carbon emissions estimated to be 7.5 Gt C yr^{-1}.

- Agriculture, forestry and land use change is reported to be directly responsible for approximately 18–24% of total anthropogenic GHG emissions each year.

- Conversion of natural ecosystems to managed systems is reported to deplete SOC stocks by an average of 60% in temperate regions, and up to 75% in the worst affected regions of the tropics, accounting for losses of up to 80 t C ha^{-1}.

Source: Keenor, S. G., Rodrigues, A. F., Mao, L., Latawiec, A. E., Harwood, A. R., & Reid, B. J. (2021). Royal Society Open Science, 8(4), 202305.

The message in these snippets of evidence is that soils used for agriculture release carbon into the atmosphere—equivalent to a quarter of what the atmosphere currently holds—and are less able to support plant growth. What the researchers normatively call damaged. Carbon loss from soil contributes to the current rapid climate change and confounds what could be done to mitigate further change.

All this is a big deal, and we explain the details in our *Mindful Sceptic Guide to the Soil Carbon Debt* because it needs more attention than we can give it here. The impact of soil degradation on climate change is dramatic, but it's not even the most critical issue. That honour goes to the consequences for soil function of losing carbon—what happens to processes within the soil when soil is depleted in organic carbon below the natural levels.

Here is the list of features that appear when carbon is lost, and it could be better news for the plants or the people and their livestock that eat them:

- reduced nutrient provision
- water retention and availability declines
- loss of below- and above-ground biodiversity
- soils become more vulnerable to erosion
- accelerated processes of desertification in drier regions
- decrease in crop yields

In short, soils depleted in carbon are less productive and biologically active than they could be. Alright, Debbie, that is enough. We get it. Soils can be damaged; when they are, they lose carbon, which is not good. But wait, there is more. Although abundant and long-lasting, **soil resources are non-renewable** on a human (anthropogenic) timescale. If soils are damaged, they take time to recover. And if the soil is lost through erosion, it can take millennia to replace. Consequently, poor soil management and loss of soil carbon are resulting in several aggregate consequences that include these:

- exacerbated topsoil losses up to 10–40 times greater than natural replacement rates—in the USA, topsoil loss rates are roughly 10 times that of replacement; while in India and China, loss rates exceed 30–40 times the natural replacement
- impairment of society's ability to grow sufficient crops
- decreases in crop productivity by 0.3% per year, expected to

aggregate to an average of 10% reduction in yields by 2050

- across the European Union, 45% of agricultural soils are considered impaired or very impaired in SOC content

In timeframes that matter to humans and their grandchildren, the soil is a vital non-renewable resource degrading rapidly, affecting current and future food production.

The financial cost of soil degradation

Most people know little about soil, but we all know about money. Gathering all this bad news and describing it as a financial cost that might be more understandable to the average Joe and the muppets he votes for is tempting. Here is what the researchers estimate.

- the cost of soil degradation in the EU exceeds €38 billion yr^{-1}, with associated crop loss costing more than €1.25 billion yr^{-1}
- in the USA, soil degradation and reductions in soil carbon are estimated to cost at least US $44 billion yr^{-1}
- globally, the effects of soil degradation compound to a total estimated cost of approximately US $300 billion each year

Source: Keenor, S. G., Rodrigues, A. F., Mao, L., Latawiec, A. E., Harwood, A. R., & Reid, B. J. (2021). Capturing a soil carbon economy. *Royal Society open science*, *8*(4), 202305.

These costs are not on the books of the farmers or the various players in the food supply chain. They are in the future and are only ever recorded as opportunity costs that rarely, if ever, appear on balance sheets. This is a problem that all natural resource users face. If a cost is not booked on a balance sheet, the business exploiting the resource dodges the costs and looks healthier. The problem of damaged soil with depleted function and ongoing degradation is kicked down the road for a balance sheet in the future when the depleted soil results in lower yield and risk of crop failure. Take the profit today and worry about tomorrow's profit later.

Hopefully, this is an ah-ha moment for the mindful sceptic. Making money today always wins over money made in the future, significantly if today's externalities compromise that future profit. This is how humans have always used resources. Benefit today, worry about the future later and let nature absorb the externalities.

Only soil is the fundamental natural resource essential to human survival; without it, we starve. It might be expedient to make a profit or even a living off the land, but that soil has to feed all the generations to come. If it is exhausted or abused now, a reckoning will make Debbie Downer sound like Matthieu Ricard, a Tibetan Buddhist monk from France called 'the world's happiest man'.

Changes from soil degradation

The sobering facts about soil degradation are not in the self-help books and don't feature in the social media algorithms but are essential to know, even for a Monk living on a mountaintop. In the spirit of self-help, I bring you some more research gems and take a deeper *Mindful Sceptic* contemplation of their confronting claims.

In the last 150 years, more than half of all soils have been damaged

WWF 2018 Soil Erosion and Degradation

Half is significant. Measured as area, it amounts to 50 million km^2, which is half the habitable area of the earth—only the livable space includes 39 million km^2 of forests that are, presumably, not damaged that much. So 50% is due to the impacts on soil from agricultural practices, especially on the 11 million km^2 of arable land.

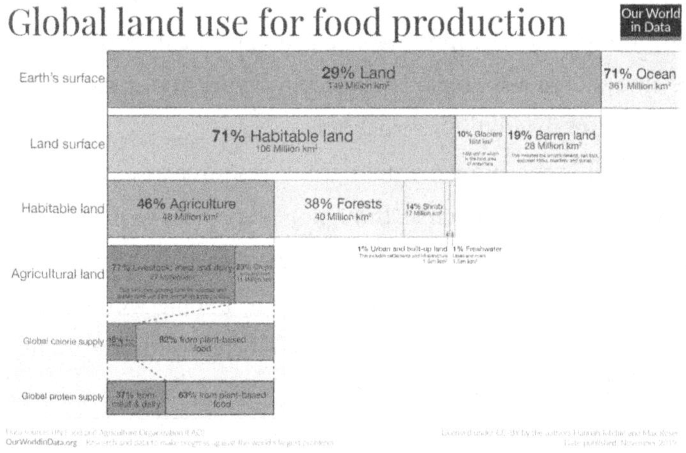

The 'damage ' could be just the consequences of agriculture and not how agriculture is practised, and it means something different from 'degradation'; otherwise, the FAO would have claimed a half and not a fifth. Damage is a green NGO's word to describe human actions. It sits nicely alongside the WWF website's impacts and threats. A mindful sceptic recognises these as normative terms that reflect values, not objectivity. The WWF does not say the soil has changed; they say it is damaged. How do you damage soil? Agriculture uses soil to grow plants that otherwise would not be growing in that soil. And when anything is used, it wears and tears, often gradually. It is this initial loss from use that WWF references. However, soil is degraded over time through loss of organic carbon, loss of structure, reduced biological activity, poorer water balance, and reduced ability to support plant growth. This happens because the soil is physically disturbed by machines that plough, plant, fertilise and harvest; also, chemically where, fertilisers, herbicides and pesticides mess with the soil biology. Crucially, soil is depleted by the removal of plant biomass. We call it harvest and even have religious festivals to celebrate the event. At the end of each cropping cycle and periodically with livestock production, the most nutritious part of the plant leaves the field to be eaten elsewhere. Over time, this is a significant loss of biomass and nutrients that would otherwise feed the soil.

Farmers know this and add fertiliser, water and pesticides to keep their crops growing. Some will even add manure and other organic sources. Damage is not the correct word, but when humans use soil for agriculture, we change the capability and condition of the soil.

When contemplating one half, there is always the half not altered. Some soils in the world are still near their natural state, and we can learn from how they function and how not to damage them. Then comes the critical fact of 150 years. A century and a half is a blink in evolutionary time and not very long in ecological time, usually measured in hundreds of years. It is a force trauma, a near-immediate impact with lasting consequences. Most of what WWF calls damage came about since the advent of internal combustion engine tractors, with Henry Ford's Fordson tractors in the vanguard during the 1920s.

Half is a lot, and it happened in an instant. Ponder these two facts for a few breaths, and the magnitude of the challenge of human survival becomes much more evident.

Degradation of soil has been accompanied by the attrition of greater than 50% of the soil organic carbon (SOC) stock in some cultivated soils, with over 2 billion ha affected globally.

Lal R. 2004 Soil carbon sequestration impacts on global climate change and food security. Science 304, 1623–1627.

This is the nerdy way to say that agriculture mines soil carbon and that this carbon loss releases carbon dioxide into the atmosphere and results in soil degradation. Half the soil carbon content across 20 million km^2 of land is no longer in the soil. This happens because when native vegetation is cleared for crops or grazing, the carbon cycle established over millennia from the interplay between the soil, climate and vegetation is disrupted. Soil is exposed to the sun and the wind; soil structure is impacted as the trees are ripped up, the ploughs level the land, and water balance changes with all sorts of consequences for the biology and chemistry of the soil. I could repeat what we have already discussed, but for contemplation purposes, here is the crux. Loss of soil carbon changes soil condition and reduces soil capability compared to its potential. Soil becomes **less good at being soil.**

Soils subjected to degradation become a significant emission source of CO_2 to the atmosphere. Soil degradation has liberated an estimated 176 Gt of soil carbon globally; a significant quotient when contextualised against the 800 Gt C held in the atmosphere.

Janzen H. 2004 Carbon cycling in earth systems—a soil science perspective. Agric. Ecosyst. Environ. 104, 399–417.

If you were wondering where the soil carbon went from clearing vegetation and soil exposure through the cropping cycle, it went up into the sky. Yep, most of that soil carbon hitched itself to oxygen and became CO_2, carbon dioxide, a gas that makes up about 0.04% of the atmosphere. It might be a trace gas, but as everyone knows, it absorbs infrared radiation and acts like a blanket to warm the Earth. It is a greenhouse gas. Given the nominal global warming potential of 1, CO_2 also provides a standard scale for measuring the climate effects of different gases. There is more carbon in the soil than in the atmosphere, but soil degradation from human land use has released roughly 22%, nearly a quarter, of the CO_2 in the atmosphere, almost all of it in the last 100 years since the advent of modern agriculture. A quarter is significant, too.

> *Averaged over the last 150 years, the soil carbon loss rate equates to 1.6 ± 0.8 Gt C yr−1. In context, anthropogenic global carbon emissions in 2000 were estimated to be 7.5 Gt C yr−1 (i.e. the rate of annual SOC loss is approx. 20% of this value).*
>
> Van Vuuren DP et al. 2011 The representative concentration pathways: an overview. Clim. Change 109, 5.

Contemplate the 176 Gt C released into the atmosphere as CO_2 from soils over 100 years. And then realise that CO_2 concentration in the atmosphere is double what it was before the industrial revolution. So when the climate activists tell you climate change is a fossil fuel problem, they are correct—the latest global CO_2 emissions for 2020 from energy (roughly 70% of all emissions) were 8.6 Gt C. But then, here is the thing.

> *Agriculture, forestry and land use change is reported to be directly responsible for approximately 18–24% of total anthropogenic GHG emission each year.*
>
> Friedlingstein P et al. 2019 Global carbon budget 2019. Earth Syst. Sci. Data 11, 1783–1838.

Agriculture is a greenhouse gas emitter. It's not easy to hear and never uttered by a politician for fear of the trolls and the media roastings that would herald the voter backlash. The reality is that agriculture uses half the habitable land and cannot stop. Agriculture must produce 22 trillion calories daily to keep people alive. Failure to meet this ongoing demand means people will starve, and children will die or live with debilitating malnutrition. So climate change is an agriculture and soil degradation thing, too. It is not politically correct to point out the emission profile of agriculture except for those who want us to give up meat, but it exists. Land clearing and intensive crop production are as significant as cows belching methane. As you have discovered, mindful sceptic curiosity and critical thinking can be painful.

> *This conversion of natural ecosystems to managed systems is reported to deplete SOC stocks by an average of 60% in temperate regions, and up to 75% in the worst affected regions of the tropics, accounting for losses of up to 80 t C ha−1.*
>
> Lal R. 2004 Soil carbon sequestration impacts on global climate change and food security. Science 304, 1623–1627.

All the contemplating completed up to this point is exhausting; you need a latte from the coffee cart that just pulled into the car park behind the tree. The barista is proudly spruiking a Tanzania Peaberry Coffee that just won an award, and sure enough, the Arabica bean delivers medium body and delightful fruit-toned acidity. Coffee is Tanzania's most significant export crop at around 40,000 metric tons annually. The Tanzania Coffee Board planned to double production in the country from 2011 to 2021 but missed its 2017 target by 41% due to decreasing yields. In most regions of Tanzania, a coffee tree becomes economically unprofitable after 20-25 years, and most of the 240 million coffee trees around the country are older. As the coffee trees aged past their productive lives, intercropping with bananas and other crops caused soil degradation. This coffee anecdote brings inequity into the contemplation. Where the people are poorest and the need greatest, the soils are generally more deficient and more prone to carbon loss. This is more than a coincidence. The rich, fertile soils of Europe, North America and parts of China contributed to productive agriculture in these regions, forming the foundation for their ongoing economic success. Equally, poor soils are a big part of why famine and food insecurity are real in Africa. This is a rabbit hole of contemplation. Once it becomes about people, it is deep and frequently dark.

Your first thought as a mindful sceptic is always, is this true? What evidence is there for these claims of soil degradation? The science quotes in this chapter come primarily from publicly accessible, peer-reviewed evidence synthesis papers that quote hundreds of sources and are written by scientists from reputable institutions—research by leading soil scientists (including Ratten Lal, Pete Smith, Cornelia Rumpel, Peter Alexander etc) as well as international agencies (FAO, Climate Change Authority). It stacks up well on the Mindful Sceptic Evidence Hierarchy, given it comes from systematic reviews, experiments, and detailed modelling.

Mindful Sceptics

Evidence Hierarchy

1. **Systematic reviews**
2. Randomised trials and multi-site experiments
3. **Natural experiments and time series of observation**
4. **Ecological process models and spatial models**
5. Before and after studies, laboratory studies
6. Case studies
7. **Expert opinion**
8. Observations

There is some normative language—degradation, damage—but most evidence is presented objectively. Next, you cannot leave a number alone, so some context is needed for the evidence.

- 2 billion ha is 20 million km^2 or a little over half the agricultural land on earth
- 176 Gt is 22% of the carbon in the atmosphere
- soil carbon loss rate of 1.6 ± 0.8 Gt C yr^{-1} is between 11-32% of the total anthropogenic global carbon emissions
- €38 billion is 7% of the EU's agricultural industry production in 2022, as estimated by the European Commission

- $44 billion is 27% of the economic output of America's farms (0.7% of U.S. GDP) as estimated by the US Department of Agriculture for 2021.

- $300 billion is 7% of the FAO estimate of global gross production value from agriculture

Placing the numbers into context tells us that soil degradation is significant for carbon emissions and most economic comparisons. However, 7% is within the margin of error and certainly within the interannual variability in production volume and commodity prices. A Nobel laureate economist concluded that the 0.7% of GDP meant we didn't have to worry. The sceptic's conclusion is much more intelligent. She will say that soil degradation is a big deal for the farmer, less so for today's global economy, but critical for long-term persistence. The mindful sceptic who gets the ah-ha moment on profit or realises that intensive agriculture has made food production an energy sink is worried by these numbers, 'stay awake a night' terrifying.

Soil degradation is not the only grand challenge for society. It might not even be the most immediate when there is geopolitical instability and a worldwide inventory of over 13,000 nuclear weapons. It certainly has one of the lowest public profiles, quickly overshadowed by the cost of living, energy crises, climate catastrophes, and war.

But ignore the research evidence in this chapter at your peril. Talk about it with your friends and ensure it gets onto your list of issues come election time. Because even if you live in a country where supermarket shelves are stocked, and the family food budget is less than 20% of disposable income, not every human is so blessed. Hundreds of millions and soon-to-be billions of people are exposed to the consequences of degraded soils on the food supply. Many are already food insecure, and as the pandemic showed, hundreds of millions more can join them in a hurry. What would you do if you couldn't find food for your family? You would likely do what it takes. Hungry people will move to find food until they are stopped or herded into camps. It could get ugly. We owe it to them and ourselves to reverse the damage to soils and ensure enough food for everyone for a long time. It is crucial to appreciate that although they appear abundant and long-lasting, soil resources are non-renewable on an anthropogenic timescale and can degrade from misuse. And where soil degradation rates outpace biogenic and geological soil replacement and recovery rates, the resource is no longer sustainable.

Globally, poor soil management and loss of soil carbon have exacerbated topsoil losses to a point where they are 10–40 times greater than natural replacement rates. In the USA, topsoil loss rates are roughly ten times that of replacement. While in India and China, loss rates exceed 30–40 times natural replacement.

Failures in soil management that lead to soil degradation will decrease crop yield and impair society's ability to grow sufficient crops. Crop productivity loss of 0.3% per year is reported from degraded soil carbon stocks, a decrease that could aggregate to an average of 10% reduction in yields by 2050—with the worst affected regions experiencing up to 50% yield reductions. Across the European Union, 45% of agricultural soils are considered impaired or very impaired in soil carbon content. And so it goes.

Key Points

- Half of the world's soils have been impacted by human activity over the past 150 years, primarily through soil organic carbon (SOC) stock loss, impacting soil health and functionality.
- Soil degradation reduces nutrient provision to plant roots, compromises water retention, results in biodiversity loss, increases erosion vulnerability, and decreases crop yields, collectively impacting soil health intricately linked to food security, environmental stability, and overall sustainability.
- Soil degradation also releases substantial amounts of CO_2 into the atmosphere, contributing to global warming.
- The financial costs of soil degradation from decreased crop yields and compromised agricultural productivity accumulate over time, posing a significant challenge for future generations.
- Soil degradation affects agricultural regions and areas with vulnerable populations, increasing the risk of food scarcity and population displacement.
- The issue is urgent and requires attention in conversations, education, and advocacy efforts to reverse the damage and promote sustainable soil management practices.

In the next chapter

As you sit under the tree next to the coffee cart contemplating these scary facts that make the consequences of climate change seem balmy...

Breathe. Humanity is so vast and voracious that any sane person is overwhelmed by our impact... Breathe again. Now, imagine how to slow everything down and make repairs. There is a way, and we come to that in the next chapter.

Chapter Seven

Can agroecology fix soil degradation?

The 12,000-year history of agriculture is all about production within nature's limits. Then fossil fuels came along 150 years ago to make us want to go back to the future.

We need a fix

The weird thing about the problem of soil degradation is that we know why and how it happens. This knowledge means that we also know how to reverse it when it occurs and, much better, prevent it from happening in the first place. When I say 'we', I mean that there is recorded evidence and, therefore, human knowledge that can reliably explain how soil is depleted, typically by the outputs of nutrients and carbon being greater than the inputs, and why this happens, growing food for profit or subsistence.

As with most human knowledge, there is nuance and technical depth that only a few people grasp: the farmer for his fields, the researcher studying the fractionation of nitrogen compounds in clay soils, the sales rep working for the fertiliser company, and the occasional advisor to the agriculture minister. And even though most of us are not wise to the subject, there is no collective excuse for allowing the problem to develop or persist. Because if it does, we are all in trouble. When people don't have access to good food, they panic. So, keeping all 8 billion of us well-fed makes good sense to retain any semblance of a stable society. Feeding people for a long time with healthy food without pillaging the environment is also a fundamental requirement to buy humanity the time and emotional space to fix our myriad of other problems. All this is especially odd when there is a ready-made solution to maintaining soil in good condition whilst growing food for the many. It's called agroecology.

> *We believe agroecology offers the way to support healthy diets and provide sustainable and just food systems to meet the needs of a growing planet — without sacrificing the planet*
>
> Doreen Robinson, Head of Biodiversity Land Management, UNEP

What is agroecology?

A simple definition from the Soils Association, a UK charity and organic certification body, is that ***agroecology is sustainable farming that works with nature.*** Recalling that ecology studies relationships between plants, animals, people, and their environment, alongside the balance between these relationships. This makes agroecology the application of ecological concepts and principles to farming. This is not just some lefty, greeny horse manure. Here is what the FAO, a United Nations outfit, says about it

> *Agroecology is an integrated approach that simultaneously applies ecological and social concepts and principles to the design and management of food and agricultural systems. It seeks to optimize the interactions between plants, animals, humans and the environment while taking into consideration the social aspects that need to be addressed for a sustainable and fair food system.*

There is a consistent message here—apply the principles of nature to growing food and use the science of ecology to help achieve it. FAO also identify in a report entitled *The 10 Elements Of Agroecology Guiding The Transition To Sustainable Food And Agricultural Systems* the aspects and benefits of agroecology that would get us to a food nirvana of feeding everyone well and meet some of the targets in the Sustainable Development Goals. The ten elements are: diversity, synergies, efficiency, resilience, recycling, co-creation and sharing of knowledge, human and social values, culture and food traditions, responsible governance, circular and solidarity economy.

Well, there you have it: a simple and easy route to food nirvana, technobabble style. The interesting thing is that stripped of the jargon and jingoism of modern times, applying the principles of nature to farming is what humans did for thousands of years. The early farmers had no choice; they had to work with nature.

Back to the future

In the famous movie, Marty McFly accidentally travels back to 1955 in a time-travelling DeLorean automobile built by his friend, the eccentric scientist "Doc" Brown. Back in time, Marty inadvertently prevents his future parents' meeting—threatening his existence—and is forced to reconcile the pair and somehow get back to the future. Agroecology is another time-travelling sci-fi trope when stripped of its unnecessary UN bells and whistles. Working with nature to grow food is precisely what the first farmers did 12,000 years ago. Let's describe what they did with the jargon thrown in.

In the fertile crescent, for example, people harnessed the high **diversity** of plant species with tiny seeds that could self-pollinate and be cross-pollinated, which made them good candidates for early experiments in cultivation.

They combined these plants to create annual and perennial crops, livestock and aquatic animals, trees, soils, and water on the land close to the rivers that became the new farms. Without knowing it, these early farmers created **synergies** of nutrient and water use that were **efficient** enough for them to stay put all year round.

Staying put was a highly successful outcome that allowed new ideas like ownership, trade and permanent structures to persist. We can safely assume that the new invention of agriculture was **resilient,** given there are still farms across the crescent today.

Archaeological researchers have used stable isotope analysis to establish that these early farmers intensified production through water management and used manure to enhance crop yields. No doubt they **recycled** plant as well as animal waste.

They may have **cooperated** beyond the innate level of comradeship essential to the survival of the hunter-gatherer groups that preceded them. The initial assumption that these early farmers were a homogenous group that traded and intermingled was questioned by genetic research from 2016, suggesting that multiple groups of people in the Fertile Crescent started agriculture but were isolated from each other. Cooperative or not, we know that agriculture spread from a handful of origin sites to every continent both by the slow migration of farmers and by transferring agricultural techniques to hunter-gatherers.

The early years of agriculture were also when **human and social values** and **culture** developed rapidly. The pyramids and the acropolis were only possible because the people had the time and energy to build them, along with the social system and **governance** to encourage or force them into construction work.

The only one of the ten elements on the FAO list that we cannot be sure was happening 12,000 years ago among the first farmers was a **circular and solidarity economy**. We assume that any early agricultural economy was a form of barter over time and attention because money came later. We know that accounting records date back more than 7,000 years in the region of the Fertile Crescent, and documents from ancient Mesopotamia show lists of expenditures and goods received and traded. Once humans had agriculture, they had an economy.

Modern agriculture

Today's intensive agriculture is a just-in-time production system, highly dependent on inputs (energy, nutrients, water) not wholly controlled by the farmer, who must also cope with a fickle market for produce. Modern agriculture is productive, yielding way more than early farmers could have imagined, but with low resilience and susceptibility to climate and market disruption. More important is why the inputs are needed in the first place. Farming is supported less and less by the nutrient-sharing ability of soil—less by nature and more by management supplying energy and nutrient subsidies.

Imagine what happens to soil when plants are grown year after year for thousands of years, and the highly nutritious parts of the plant are harvested, eaten away from the farm and defecated into the ocean. Soils have had their nutrients and organic matter mined. This is an uncomfortable truth. Human and social values have changed since the first farmers experimented with crops and livestock. Today, we have corporations with capital and technology to farm intensively at scale. This must happen to feed most of us who live in urban areas and can't feed ourselves because it successfully converts ancient energy into food. And humanity cannot yet live without it.

What agroecology offers is a way back to using the soil as a natural resource rather than mining it. Unfortunately, agroecology is not **the way**, at least not yet. But there is great hope, especially from the stories of a few courageous pioneers.

Eccentric farming pioneers

We laugh at the antics of people who are different, odd, out of the box. It is how we cope with our fear of the unknown and unusual. The thing is that we shouldn't laugh at the strange. Humans will need unusual farming techniques when the familiar ones start to falter. We will also need ideas for regenerating degraded landscapes as the soils lose their potency and fossil fuel inputs become scarce and expensive. So, I would like to introduce you to three eccentric farming pioneers, a selection from the many brave souls who not only dared to be different but also told people about it.

Alan Savory and Holistic Management

Many years ago, I met Alan Savory, the Zimbabwean inventor and passionate advocate for holistic management, a form of farming that integrates agricultural production into nature and the farming people by understanding complexity. He was white-haired and earnest with a firm handshake. I doubt he would remember me, but I will never forget his intensity and furrowed brow. Alan's ideas have spread far and wide through the Savory Institute, which promotes the mindset of making complex decisions in food production from an ecological perspective. They teach a suite of planning procedures under the banner of Holistic Management, including planned grazing, land planning, financial planning, and ecological monitoring. However, the real lesson is that food comes from nature—a complex web of interacting organisms. Ecological processes drive food production.

In his book *Holistic Management: A Commonsense Revolution to Restore Our Environment,* Alan explains how his insights came from tracking animals across the savannas of southern Africa. He saw how grazing animals gathered in herds to reduce the risk to each individual of ending up in the jaws of a predator and that, in turn, the predators tracked and moved the herds across the landscape. Not herded, though, because such a concentration of herbivory had to move; otherwise, all the plants would be eaten. So animals trekked across the grasslands and open woodlands, sometimes in great migrations that followed the rain, but everywhere herds moved, grazing the vegetation in pulses. This is the opposite of the way modern humans rear livestock on rangeland. Ranchers let animals roam across large fields that end in a fence. In these constrained conditions, they eat a little and often all the time—a continuous grazing system. Alan's insight was to mimic the great herds in livestock production systems. His cows moved as a single large herd across the landscape, eating and defecating a lot, leaving the plants time to recover when the herd was elsewhere.

I recognised where Alan came from, having spent two years in Zimbabwe and another seven in Botswana. I saw the ungulate herds followed by the lion prides and walked over the land 5,000 buffalo had just mauled. It is so apparent when you see it. My ecological background meant that I immediately knew how Alan came to understand the process of pulsed grazing and 'regeneration with rest' that characterises many savanna ecosystems and helps make them so productive.

Alan is a purposeful presenter and engaging enough to have his TED talk on *How to green the world's deserts and reverse climate change* reach close to 6 million views on YouTube. But the reality is that no matter how logical and profitable holistic management can be for many livestock and mixed production systems worldwide, it has little traction. It is still in the backwaters of agricultural thinking, well away from the mainstream. Part of this is because Alan Savory is unusual. The average Joe is wary, and maybe a little jealous, of a dude who runs barefoot in Africa amongst the lions and the buffalo.

Peter Andrews and Natural Sequence farming

Australia is an old continent famed for drought, fire and flood. After living in NSW for nearly 30 years, I can confirm this weather-related notoriety is justified. I have seen the wildfire dispense with the fence at the bottom of my garden and bailed the water from the shed after a flood. Australia is a tough place to grow food. The weather is an obvious challenge, but the soils are old, too. They have been exchanging nutrients with plants for millions of years and have lost much of their innate capacities and organic matter.

On my travels, I have seen the dry grass, the thirsty trees and the parched soil stretch for mile after mile. Farmer after farmer works all hours to make the average 4,000 ha farm productive. They need a lot of land to make it work. Their European colleagues make a living on an average of less than 100 ha. Then, out of nowhere, an emerald green paddock will appear on the horizon. The first thought is that a farmer has invested in inorganic fertiliser or has tapped the groundwater to irrigate a crop. As I get closer, it's clear that the greenery follows the drainage lines in vast swathes. Chances are, I have stumbled across a property managed for natural sequence farming, the intelligent idea of eccentric farming pioneer Peter Andrews. The simple and elegant concept is to regenerate a degraded landscape by slowing the water flow with leaky weirs and judicious planting of whatever plants will grow to help hold the water and build the organic carbon in the soil. These plants might even be weeds or alien species. It only matters that they hold on to the water.

Enthralled by his idea and initial success on his farm, Peter Andrews wrote two books, *Back from the Brink* and *Beyond the Brink*, about how to recover the dry landscapes of Australia by holding water in the landscape. What Peter Andrews figured out was that plants would help soil absorb the water that often fell in torrents, hold it back, higher up slopes, to release it more slowly. The more plants, the better the sponge. It didn't matter what type. It also didn't matter if a few physical barriers were put in to assist. He wanted to mimic nature, not recreate it. As an advocate, he earned the ear of senior politicians, ran training courses, and was awarded the Order of Australia Medal. Research was commissioned to confirm that his techniques met the claims, curiously by hydrologists rather than ecologists, and as with all things involving nature, the results were equivocal. A few devotees follow the methods with the support of the TALS Institute, where Peter Andrews is chair. And yet, two decades after his 2006 book, natural sequence farming remains at the margins.

Part of this is because Peter Andrews is unusual. His idea works, but after its initial regeneration, his farm fails, and his capricious personality puts off many would-be advocates.

Ernst Götsch and Syntropic Farming

I have not met Swiss farmer **Ernst Götsch**, who developed techniques that reconcile agricultural production with landscape regeneration through principles and practices globally known as Syntropic Farming, but I would like to. His idea is to change the way we see, interpret and relate to the farm so that food production becomes cooperative with nature rather than one of dominion over it.

In Ernst Götsch's vision, holes become nests, and seeds become genes, weeding becomes harvesting, the competition gives way to cooperation, and pests and diseases are seen as the *"agents from the department of optimisation of life processes"*. His 410-hectare farm in Brazil is regenerated from degraded land with the resurgence of 14 springs and the reappearance of native animals. It is a testament to the concept.

But you may not have heard of syntropic farming; my spellchecker hasn't, and I hadn't until recently, but Ernst Götsch has been at it for 40 years, and his nature-based techniques for agroforestry are still not mainstream in that niche area of food production. Part of this is because Ernst Götsch is unusual. He may be a normal enough fellow, but he speaks a very different language to the average farmer. Succession, companion planting, and pruning are all tied up with trial and error.

These words are not what farmers talk about in the pub once they finish complaining about the footy.

Unusual agricultural techniques

Savory, Andrews and Götsch are pioneers, but they shouldn't be. They share the challenge of doing different and counterintuitive to a farming industry and society that has grown fat on a command and control model of production that relies on energy subsidies. What they have done is understand how plants grow and the ecological processes that make it happen. They figured out that it is soil replete with organic carbon, moisture and nutrients that must be maintained for a farm to be productive, then they worked with their circumstances—grazing for Alan, water for Peter, and trees for Ernst—to find the way.

They are also a little odd, eccentric even. I admire them for this because, being a little unorthodox in my views on food, ecology and diet, it takes a lot to go against convention. You must believe in your contrary view because there will be doubters, critics and trolls. Most people are put off by the eccentric, a little scared by their passion for what is, for now, the extreme.

The thing is this. Intensive agriculture is a massive energy sink that persists because of fossil fuels. A change is coming because the ancient energy pulse will end. This is the truth: a resource constraint on how we grow food to meet the 22 trillion a day challenge. Oh yes, and the climate is changing, too, which makes matters worse. As we have been learning, the soil that supports all the plant growth is degraded by intensive agricultural practices.

Holistic management in agriculture and regenerating degraded landscapes through natural sequence farming and syntropic farming principles are ways forward from this over-reliance on inputs. They are similar in their understanding and use of ecology and ecological principles to grow food where the energy is innate, and nutrients are cycled through complex interactions between organisms and their environment—livestock are often key to agroecology.

Savory, Andrews and Götsch are our visionaries who see what is possible when the oil runs out. A return to production that is more ecological, circular and potentially sustainable.

Wherever possible, farms should revert to ecological principles to support production with simple measures such as retaining vegetation cover on the soil, reducing tillage and recycling as much organic matter as possible within the farm. The benefits of agroecology are the circularity of nutrients that make food production less dependent on inputs. This is essential given that the inputs will be squeezed. But if we do this everywhere and ignore intensive agriculture, it will not feed the 5 billion people expected to live in urban areas by 2050. Agroecology offers **a way** for part of the food production system. In time, it could become the dominant and most widespread mode of production because it is less reliant on commercial fertilizers and pesticides and energy use.

But it is not a silver bullet, for they do not exist.

Existential threats to farming

People like living together. The UN reports that the world's population in urban areas is 58%, projected to reach 68% by 2050. That would be 6.6 billion people who obtain food security from supermarkets, fast food outlets and corner stores. As half the people live in urban and built-up areas to take advantage of the economic opportunity of aggregation—just 1% of the global land area —it leaves the farmers to look after the bulk of the land. For example, Australian farmers manage up to 60% of the country's landmass and use 70% of its diverted freshwater extractions to grow crops and raise animals to make a living. These farmers use soil and water to grow crops, raise animals and tend not to think about existential threats. This is good news for urban and suburban populations worldwide who don't have enough land or the skills to grow their food.

Not all farmers are managing the land to grow food for sale. Globally, most of the 500 million farmers grow crops and raise animals for themselves and their families. FAO surveys show that five of every six farms in the world consist of less than two hectares, putting most farmers into this subsistence category. The remaining 20% of farms have to feed everyone else. And they have for a long time. Localised pre-industrial collapse aside, farmers have fed the human population as it has grown slowly and then rapidly, thanks to the intensive input-driven farming systems we are familiar with today. Humanity asks that they keep doing it. We all have to eat and are unlikely ever to have a realistic alternative to soil-based food production to grow the food to feed 8 billion people and their pets. This task is about to get more complex and production riskier.

Australian rural journalist Gabrielle Chan suggests that "*farming both contributes to and is endangered by the biggest existential threats of our time*": climate change, water shortages, soil loss, energy production, natural disasters, zoonotic diseases, population displacement, and geopolitical trade wars.

As mindful sceptics, we would add the following to this already scary list: population growth, urbanisation, soil degradation, peak phosphorus, just-in-time supply chain fragility, declining insect populations, inflation, and Uncle Tom Cobley. The existential threats to farming are many and cover a vast economic, social and ecological territory. As a Mindful Sceptic, you can see the complexity of modern food systems and recognise their fragility. Feeding everyone well needs awareness and understanding of dozens of subjects, skills and psychologies just to cover the basics of sustainable food. It's complex.

Existential threats to farming barely reach the farmer. They are too busy running a business. They have to be financially astute, managers of labour and equipment, practical MacGyvers able to fix a combine with a bit of wire, and expert meteorologists. Each day on the farm will bring a hundred challenges, needing a decision. It is cruel to expect farmers to be across all the practicalities and be aware of the threats happening beyond the farm gate. Farmers should get help for the existential threats to farming, especially from public policy. Politicians need to get their heads around the policy options and be supported by a public service resourced and skilled for the purpose.

Here is Gabrielle Chan again.

> *...there is a barrage of contradictory policies around food growing, and no Australian political party is doing serious thinking about how to knit together food, farming and environmental policies to continue feeding the population while mitigating climate change and biodiversity loss*
>
> Gabrielle Chan, Guardian Australia's rural and regional editor

Tim Long says the same thing about the UK in his excellent book *Feeding Britain*. He suggests that food policy in the UK is wayward or nonexistent, which is not helping. And for the UK, it is a huge risk given that the country still imports half its food supply from overseas. I fear for them when their money runs out, fiat currency notwithstanding.

Food, farming and environmental policies

It should be easy enough for the politicians. All voters and their kids must eat and spend between 10 and 50% of their income on access to food. For billions worldwide, keeping the family fed is a daily challenge. Even for the wealthy, there are questions about what to eat and how to cook it. Everyone is thinking about food.

Policy is not so clear-cut. Food policy is about what food does in the economy and society. It covers the price of food, the guidelines for which food is healthy, and how to maintain food safety. In some countries, food policy may extend to issues of food security that may cover supply and access. Food policy is about people and how to keep them from complaining about foundational human needs.

Agricultural policy is different, at least in the mature economies. It tends to focus on farming and the people who farm. Invariably it is about ensuring the local food production systems are viable, ongoing businesses. Policies are implemented to maintain commodity prices, support trade and maintain a rural labour force. More often than not, these policies pull the subsidy lever. Agricultural policy is about using land and increasing the commercial use of soil resources.

Environmental policies are different again. They are loosely grouped into environmental protection, which covers what not to do and the penalties for polluting the environment, and conservation, which is about what to protect. These policies are often in conflict with agriculture because the truth is that there is a trade-off in what can be done with land and with nature before there are consequences. Environmental policies are generally about values.

There is no modern democracy where a single minister or bureaucracy handles all three of these policy areas. Typically, they are in different ministerial portfolios and civil departments. But I digress. It's easy to do in this complex arena of food and people. And as all mindful sceptics know, it is not a bad thing as insights often come from tangents, but we should return to the soil.

What about soil degradation?

Our eccentric farming pioneers figured out that the ecology of soil mattered to growing food. Each, in their own way, recognised that healthy, biologically active soil was essential to food production. Perhaps they were not explicit in communicating this fundamental, but they knew it to be true. With all its existential threats to farming, the broader human social and economic system has no time for soil. It's just dirt. Instead of a Minister for Soil backed by a line agency staffed with civil servants trained in the science and application, policies that affect soil are scattered to the winds of food, farming and environmental portfolios, if they exist at all. You must dig deep into the manifestos of major political parties to find the first mention of soil. For example, the Australian Labour Party mentions soil twice in its 65,444-word, 157-page *Labor's 2021 National Platform*—once on page 14 in a general reference to the environment

> *Australia's natural environment, with its **soils**, waterways, wildlife and coasts, is one of our most important economic assets*

and again on page 52

> *Labor will ensure national environment funding programs and other volunteer and community programs, like Land-*

> *care funding, support local sustainable agriculture,* ***soil management****, conservation and environmental priorities and contribute to national priorities to improve biodiversity and resilience".*

That is it. Nowhere else. It would seem that along with what Nate Hagens calls our energy blindness, we are soil blind, too.

A mindful sceptic knows better. She will open her eyes and learn about soil and especially the benefits of agroecology because they inform how agriculture must be practised in the future. I say 'must' because the fossil fuel energy pulse will end, perhaps sooner than we would like. Modern intensive agriculture is a net energy sink. It absorbs more inputs than it produces in food energy outputs. Unless the energy transition provides more than just electricity, farming might need to stay as one of the remaining long-term users of oil and gas reserves. Remember that most of the nitrogen that goes into grain and livestock production and about half of the nitrogen in human bodies comes from the Haber-Bosch process. This industrial fertiliser production system runs on natural gas and is crucial because eight billion people cannot live on bread alone. They will need nutrients to go with their calories. Then, there is the soil degradation that comes with the intensification of agriculture.

It is unrealistic for farmers to be across all the existential threats to farming. There are too many of them, and they cover vast intellectual territory. Farmers are going to need help to mitigate the risk. Failure of the food system is the biggest threat of all. It is a real risk with ugly consequences. If farmers fail, then so does everyone else. Agroecology recognises the circularity and the nutrient story to provide a solution to mitigate the risks and rebuild natural capacity. Feeding everyone well will need it. If that doesn't get you, ask yourself why soil is not on the agenda at election time.

Key Points

- Soil degradation is a well-documented problem driven primarily by nutrient depletion and carbon loss due to intensive agriculture. Healthy soil is essential for global food security, ecosystem stability, and climate regulation.

- Agroecology is a promising solution to combat soil degradation that works with nature, applying ecological principles to agriculture with many pioneers exploring innovative farming techniques such as syntropic farming and holistic management.

- Farmers face numerous challenges beyond the farm gate, and need support from increased awareness, understanding, and policies.

- Humanity must embrace change, support pioneers in sustainable agriculture, and adopt holistic approaches that ensure food security while mitigating the risks associated with soil degradation and other existential threats to farming.

In the next chapter

Soil degradation is an existential risk even though we know how it happens and have many ways to prevent and reverse its worst effects. It is not a science issue. Next, we have a look at the people problem.

Chapter Eight

Always tackle the social issue first

Sometimes, you read a well-meaning research article, and all is well until, boom, it hits you. The authors are from somewhere beyond the Orion nebula, over 1,300 light-years away. Here is evidence that the aliens have landed.

Humans are extraordinary

Think of any topic, detail, or weird example, however obscure, and a human has been there, done that and recorded it for posterity on Tic-Toc. Ubiquitous human curiosity means that any academic topic has been pursued into its dim recesses. I disappeared for years into the feeding behaviour of spirostreptid millipedes, tubelike creatures that eat leaf litter and burrow down into the soil for half a year. Believe it or not, that was a career move.

Farmers and the scientists who support them know what is going on when it comes to growing food and the soils where the food plants grow. This makes sense, given humans have been in agriculture for 12,000 years. We know about nutrients, soil carbon, and the impact of pests and diseases. We know how and when to irrigate crops and when to send the cattle to market. We can even predict from historic patterns and, using sophisticated soil models, how all these things will change given various disruptions, including climate change. We can 'see' the land in extraordinary detail from satellites or sensors attached to drones in real-time and then send a remote-controlled machine to deliver fertiliser or pesticides to the right spot for maximum effect. Even before all the hi-tech wizardry, farmers found ways to cope with natural climate variability, extreme weather events and wildfires. They figured out how to match management practices to environmental conditions for optimal yields, and some of them even learned how to achieve sustainable use of land resources through agroecology.

We also know when soil is degraded and why. In short, humans are extraordinary.

All this knowledge capture is handy because, along with finding 22 trillion kilocalories per day to feed everyone, the pressures on the global food system are increasing, not least from the consequences of soil degradation. In the big picture view, summarised in a review of sustainable land management for global food security, the biggest pressures on food production come from

1. growing demand for food in both quantity (kilojoules of energy) and quality (proportion of animal protein in the diet) for an expanding and wealthier global population

2. competition for productive land with other land use, especially biofuel, urban expansion and other non-food uses

3. land use and land management practices that degrade land and diminish soil health by lowering nutrient levels and organic

content;

4. global supply chains and agribusiness that drive down prices for farmers, pressuring them to farm intensively and unsustainably, and

5. impacts of anthropogenic climate change that increase year-on-year variation in yield and frequency of extreme weather that threatens the resilience of agroecosystems and the stability of food production.

So even though we know what should be done, these kinds of pressures are immediate and impact the capacity to act on what we know.

How farmers use knowledge

Farmers are courageous human beings. They take incredible financial and personal risks when investing blood, sweat, tears, and money in a cropping cycle or rearing livestock. The last thing they want is to try something because it might be helpful. They need to know it will work. Few farmers have the luxury to absorb an opportunity cost.

Farmers prefer to know that innovation will work before adopting it. Remember, arriving at the current methods has taken 12,000 years of trial and error. The novelty has been weeded out over the millennia, replaced by the tried and tested. Farmers are more comfortable with what they know. This knowledge comes from experience, local history and culture, including what their buddies tell them over a few beers in the pub. All the accumulated knowledge is both vast and local.

Farmers in the Sahel have no way of knowing that data from a drone might help their cattle. Nor do they know how to use such technical information. What they know about where to graze the cattle works. It comes from tradition passed down through the generations, and the cattle grow well when it rains, survive when it's dry and provide enough for the people to persist. It always has. Except that now all the cultural bets are off. Thanks to fossil fuel inputs and technological advances in medicine and welfare, there are many more people than before. The climate is changing, and the soil is depleted. Suddenly, innovation is essential for survival.

How to get knowledge used

When traditional agricultural practices no longer meet the needs of a local population or the farmer's bottom line, there are two options. The first is to subsidise the food and welfare needs from outside. This happens in cities all the time. Urbanites live on a food subsidy from outside the city. External support is increasing on many farms, both from policy that supports current farming practices, from trade deals to direct subsidies, and by farmers generating their means of support, such as when a family member or two takes a job outside the farm. The second is to change the land management practices to grow more food.

Scientists have become fond of suggesting ways to achieve the second option. Let's look at an example—nine ways to improve land management practices outlined in research commissioned by the World Resources Institute—with a mindful sceptic eye and my experience of more than 30 years witnessing what happens on the ground and in the corridors of power.

Strengthen knowledge management systems and access to information.

This recommendation is standard: use knowledge. The challenge is that knowledge is difficult to manage because it needs understanding and insight before implementation. Access to knowledge is easier than it has ever been. Just type anything into Google Scholar or your favourite AI chatbot, and off you go. However, skills are needed to interpret what you find—just having a database of research references is not a knowledge management system.

> *Increase communication and outreach, using champions and direct engagement with farmers.*

Talking to people is always a good thing to do, but with a value proposition, engagement is easier to maintain. Farmers are naturally cautious, and they have every right to be. Their livelihood depends on making the right calls.

> *Foster connections between communities and government and non-government institutions to enhance knowledge sharing.*

Many rural communities are not fond of governments. The government's command and control tension threatens the delicate balance in the social contract. Farmers are often libertarians with a strong will to make their own decisions. Just getting everyone in the same room is a challenge, let alone speaking the same language.

> *Establish demonstration sites on landholders' properties, and develop networks of landholders (or use existing net-*

works where applicable), to encourage cooperation and communication between landholders and provide a focal point for promoting SLM practices.

I have been to many of these properties, and they are usually great, standing out like a green thumb in the landscape. All that lush grass and a procession of government visitors can leave the farmer without any mates down the pub. As we know from any streaming service, people get jealous and catty everywhere. SLM is the acronym for Sustainable Land Management and a great way to put people off.

Support institutional and policy reforms, particularly for strengthening property rights.

I have tried to do this and have led science teams providing policy support. It is tough. Policy wonks are often subject agnostic. They are technocrats with an understanding of legislation but no idea what carbon does in soils. By the time you had explained the basics of soil health, they were asleep—one senior civil servant nodded off on me in a one-to-one presentation.

Support capacity building, particularly in community-based management of natural resources.

This one is good, notwithstanding the social contract mentioned earlier. The reality is that communities manage resources anyway. They already have capacity, so sucking eggs becomes a problem.

Increase support for integrated landscape management that brings together sectors and stakeholders from agriculture (including grazing and cropping), intensive agriculture, forestry, urban and conservation to jointly plan, design, and manage their landscapes and institutional resources for improved agricultural production, biodiversity and ecosystem conservation and sustainable livelihoods.

A few years ago, I ran a science team with more than 100 years of professional experience working with these stakeholders, and none of us had seen them come together in anything other than to argue their corner. Cooperation is the answer, but it has to be agreed upon because social capital is a fickle resource.

Reinforce economic incentives and private sector engagement.

Eventually, consumers must pay a realistic price for the commodities they enjoy. But, the economic system is designed around profit and rent—the most efficient ways to convert natural capital into cash and financial assets. Letting that system loose on natural capital got us into our unsustainable mess.

Mainstream investments to catalyse adoption of these practices as a strategic component of food security and climate change adaptation programs.

Ask Albert what he thinks.

I apologise for this little rant, and I don't intentionally single out the World Resources Institute. The list illustrates the disconnect between reality and the wonderland of policy and politics. At some level, everyone knows there are structural problems in our food systems that begin with land management, then carry on into a host of social issues that take us away from the core challenge.

The Importance of Soil Health for Humanity's Survival

Humanity as we now know it will collapse or not, depending on how we look after the soil. There are plenty of other triggers and sources of collapse than soil—not least the thousands of nuclear warheads or the end of the fossil fuel spike—but that cannot take away from the truth. If soil degradation continues to expand and our agricultural systems remain reliant on soil as a factory, then collapse is inevitable. Degraded soils cannot produce enough food to feed the population. There is a natural limit.

It will be possible to grow food in vertical gardens, make burgers from algae, and farm without fossil fuels, so there is a possible future where humans can persist without soil (assuming there are fewer people than today). The challenge is getting to that future in one piece, which means for at least another 50 to 100 years, global soils must be the source of most human food. As we have seen in this book that flies high above to see the issues and to save you time and tedium of the details, we know that global soils are in poor shape because of agriculture. It is what it is.

Here is a reminder of what happens. When natural vegetation is cleared to grow crops or rear livestock, the soil is exposed and disturbed, and its biology is disrupted. This is fine for a while as minerals are released, and soil organic matter exchanges nutrients with the roots of crop plants. But exposure is terrible for retaining organic matter, and between 50 to 70% is lost within a few years, forcing farmers to add inputs to maintain yield or avoid crop failure. After a few decades of use, most agricultural soils are below optimum capacity to produce food because their capability is compromised—the FAO estimates that 40% are far enough below the optimum to be called degraded.

This is not fiction. All farmers, big and small, know this to be true from experience. They know that healthy soil is essential and that soil can get sick. Each farmer tries their best to look after soil health and maintain their business.

Science research is just as sure. Observational data catalogue the loss consistently, and experimental comparisons with undisturbed soils back up loss rates. Then, there is detailed research on how and why soil carbon declines when soil is exposed and ploughed. Globally, the current estimate is that land-use change and soil cultivation have contributed roughly 136 gigatons of carbon (Gt C) to the atmosphere from a shift in biomass carbon since the beginning of the Industrial Revolution, with depletion of soil organic carbon (SOC) accounting for a further contribution of 78 Gt C. Annual global carbon emissions from fossil fuel use (some of it in agriculture) is 10 Gt C. The amount of carbon dioxide in the atmosphere is 874 Gt C.

Pause on these numbers for a second. What springs to mind? Maybe something like this

- 136 is roughly 16% of the current atmospheric carbon
- add the 78 from soil carbon losses, and we are at 24%
- each year, another 1% is added to the atmosphere from fossil

fuels, 10% per decade

Climate change is no surprise, given what human activities have done and continue to do globally. We are messing with more than the soil.

The loss opportunity

Sceptics with optimistic eyes will also see an opportunity in these soil loss numbers. If land management for agriculture has released so much carbon from the soil, could a management change return some of that carbon? Well, this is true. The science shows that it is possible to achieve carbon sequestration in agricultural lands through careful vegetation management. Science also tells us that this sequestration is good for climate mitigation. Sequestering organic carbon in the soil to repay the soil carbon debt could remove between 0.79 and 1.54 Gt C per year from the atmosphere. If you are super-optimistic like the FAO, then the potential is at 2.45 Gt C per year, more than the yearly CO_2 emissions from the aviation sector.

Few take this opportunity seriously. Even before the carbon offset markets took a hit from the additionality failure of some forest offset projects, fewer than 60 projects (half of them in Australia) designed to address soil organic carbon in croplands, pasture, peatlands or wetlands were registered under compliance or voluntary carbon markets, providing under 50,000 tCO_2e in removals per year globally—nothing compared to 1,500 projects covering 12 m hectares of land in the forest sector. People entrusted with national policies on emission reduction are equally cautious. As of 31 March 2020, 186 parties (185 countries plus the European Union) had communicated their first Nationally Determined Contributions (NDCs) to the UNFCCC secretariat as per the Paris Agreement. Only eight of these NDCs (4%) present targets for soil organic carbon within their intended mitigation options. A round of applause to Armenia, Burkina Faso, China, Japan, Malawi, Namibia, Uruguay and Zambia. I'm still trying to figure out what all the others were thinking.

The narrative of raising soil carbon levels reads well for broad societal outcomes—double-digit increases in yield potential, particularly on degraded lands, higher household and national food security, reduced risks from disasters, improved water quality and lower rates of displacement and migration—but it is not society that must change practices. That task rests with the farmer. Farmers and land managers need a compelling value proposition because they take risks and put in the extra effort to build soil organic carbon on top of already burdensome everyday management activities. There needs to be more than just a storyline around public good for farmers are practical folk. They need more than the promise of money from the air. Soil carbon sequestration actions will cost them money; anything from $10 per tCO2e for afforestation and conservation tillage to $50-100 for changes to continuous and rotational cropping.

What will it take to invest in soil carbon?

Several leading scientists in this field suggest that the value proposition for soil carbon has to meet five elements: 1) enhanced productivity, 2) improved risk management (for example, resilience to drought), 3) superior market access (for example, certified value chains), 4) financial returns to carbon assets, and 5) government support (for example, environmental subsidies). This list makes good sense without resolving the farmer's risk. He still has to think these elements will come together when he reduces tillage, avoids fertiliser use, keeps a cover crop or adds organic amendments to soils. Meanwhile, the forward contract for next season's grain crop is as hard-fought as ever.

The scientists suggest a priority to include soil organic carbon into existing international frameworks such as the UNCCD, UNFCCC, Ramsar, and the Global Reporting Initiative. And here I part company. International conventions and UN bodies are important to keep the global dialogue going under the premise that it is best to keep everyone in one tent. But they tend to fail on the ground because the solutions they suggest are a million miles away from the farmers and their everyday decisions. The cold reality is that anything top-down has failed because of the premise under which top-down operates—the autonomy of individual choice. Investing in soil carbon requires the farmer to decide what is the best and the right thing to do on his farm.

Whatever the innovation, tackle the social issue first.

As you can tell, I'm not too fond of the World Resources Institute's suggestions for improving land management. They are naive at best. I present them because they show how hard it is to deliver a top-down improvement to land management, however well-intentioned. A mindful sceptic would favour options for management that can leverage human psychology. Knowledge on its own will not change behaviour. Emotions change people, even if that means guilt, passion, fear, or hope. So my suggestion is to

1. use the science to show the consequences of past, current and future land management on the various values people gain from

land use and then

2. offer up compelling options for behaviour change, then

3. let the people decide for themselves

Remember, too, that worldwide, 500 million farmers are not fully commercial; they grow food to feed themselves and their families. These smallholder farms are net energy producers but only just. And yet they feed more than 2 billion people, including about 80 per cent of the food consumed in Asia and sub-Saharan Africa. For them, any change to tried and tested practices doesn't risk profit margins; it threatens livelihoods.

And we always remember what Albert Einstein so famously said: *We cannot solve our problems with the same thinking we used when we created them.*

Key Points

- Thanks to the extensive knowledge and experience humans have acquired over 12,000 years of agriculture, farmers and scientists possess a deep understanding of soil, nutrients, irrigation, and crop management.

- Pressures on food production, including growing demand for food, competition for productive land, land degradation, global supply chains, and climate change, necessitate sustainable land management practices.

- Farmers are pragmatic and prefer proven traditional methods. They adopt innovations cautiously and through incremental steps, drawing on their experience, local history, and cultural knowledge.

- Soil carbon sequestration is a means to improve soil health and mitigate climate change. Only the value proposition for soil carbon must include enhanced productivity, risk management, market access, financial returns, and government support.
- Smallholder farmers feed over 2 billion people globally and are vital to food security. Any changes in their practices can significantly impact livelihoods, making it essential to consider their perspectives.

In the next and last chapter

After a few beers in the pub discussing the many details and complexities of food, people and soil, what conclusion would an economist, a demographer and a soil scientists come to?

Chapter Nine

An economist, a demographer and a soil scientist walk into a bar

Malcolm has been an economist for a long time. He has seen the miracle of neoliberal economics feed people to the point of bursting, and whilst he has heard that there are folk who don't know where their next meal is coming from, they live far away. All they need to do is wait for the trickle-down to reach them. Food was never really on his radar, given it makes such a small contribution to GDP and is already a mature industry with the key players dominating the global market for profitable business activities, especially selling fertiliser, pesticides and the machinery to deliver them.

Stella has always been fascinated by numbers and wanted to be an actuary. Not quite making the grades to train for the pointy end of the insurance industry, she settled for demography. It was still about people and numbers, so she found the work rewarding, especially when the computers became powerful enough to make long-term projections from complex interacting variables. She knew that the algorithms relied heavily on past human behaviours, especially the numbers of births per woman and how that changed with development and wealth. She also knew that most of the models assumed that there would be enough food, so starvation was not a parameter, but thought nothing of it.

Babatunde grew up in the suburbs of Lagos. His uncanny resemblance to his maternal grandfather gave him his name and perhaps academic prowess, as his grandfather was among the first Nigerians awarded a Doctor of Science. Babatunde himself has just become a Fellow of the Soil Science Society of Nigeria after three decades of service to African soil and the farmers who till it. He knows all about the ferralitic soils of the south, the pockets of vertisols in the east and the vast swathes of ferruginous soil in the north. He can still feel the blisters from the endless treks to map the subtle changes amidst these great soil series. "I am sure it is my round", he says, standing to gather the empty glasses in clean fingers.

It was a quirk of chance that these three would ever enter the same bar, not because of geography, culture or lifestyle but because they had very little to talk to each other about. They live and work in their bubble. The narrow thinking that gets people ahead in any given discipline lowers their bandwidth for anything else. You don't get to be a soil science fellow by fiddling with economics or a professional economist without believing in *Homo economicus*—the rational man who pursues wealth for his self-interest and avoids unnecessary work by using rational judgment—and getting dirt under your fingernails. No, like has plenty to say to like but then is shy around everyone else. This is the way.

The title of this book is odd on purpose. Not many people have put food, people and soil in the same sentence, but you know now that they should. Malcolm, Stella and Babatunde have a great deal to talk about. They must understand each other's knowledge and expertise, even though Malcolm's is shallow and fragile. They need to explain to each other what they know. Doing it in a bar might lubricate the uncomfortable process.

Connections are what the Mindful Sceptic Guides are about. Seeking and understanding the multitude of links between subjects, ideas, and paradigms is a core skill of a mindful sceptic because it helps hone critical thinking and satisfies curiosity. It also adds the secret sauce we hope our pub patrons will find as they share stories over a pint of Theakston's. It is what you have found in the previous chapters. In the control and manipulation of soil to grow food at scale, the impact that intensification of agriculture has on food prices, and the flow on effect to more people.

Stella reaches for it as she leans over her plate of fries. "Surely this is just systems thinking", she said.

That's it, Stella, that's what we want, systems thinking.

> *an approach that examines the interconnectedness and relationships among components within a system, emphasising the understanding of how these parts influence each other and the entire system as a whole.*

If humanity is to meet its needs on a finite planet and not pollute itself to extinction, we need more than the perception of individual elements. We have to grasp the intricate web of relationships among those elements and comprehend the feedback loops and ripple effects where actions within a system create consequences that, in turn, influence further actions. We want the holistic perspective, the bigger picture of consequences, and that starts with talking in the pub.

Prices affect food access, which will feed back on production and people. Degraded soils will not produce as much food, affecting prices. People will reproduce or not depending on their nutrition, which is about how much food they can access. It's all a bit Blind Freddy when you think about it. The challenges come in the subtleties of these big systems. If there is a change in one part, how far does it ripple, and which interconnected elements are affected?

Systems thinking gives us a problem-solving approach that looks at the symptoms—rising global food prices, lagging density dependence in population, degrading soils—to uncover and understand underlying structures and relationships.

A mindful sceptic is an innate systems thinker who will automatically look for connections and start strategising over the evidence options before making any decision. This is the way. It takes more time and considerably more effort than the blindness and prejudice of modern populism. Still, it is far more likely to ease anxiety and foster solutions to our biggest problems. After this brief journey across food, people and soil, it is clear we can't keep flogging soil and expect to feed 8 billion people, increasing at 8,000 additional souls every hour on food they can afford.

Back in the bar, our trio are easing their way into the connections between their disciplines and lapping up the ah-ha moments. And as all 'three go into a bar' stories must end with a punchline, here it is.

An economist, a demographer, and a soil scientist walk into a bar... and never leave.

Evidence Sources

All the *Mindful Sceptic Guides* are compiled from decades of experience with the science literature, but I also gathered specific *Evidence Reviews* that you can access on the Mindful Sceptics website by clicking on or scanning the image below.

Mindful Sceptics

evidence reviews

scan or click the QR code for a deeper dive into the science sources for the Mindful Sceptic Guides

SCAN ME

Should your curiosity stretch to the spirostreptid millipedes I mentioned earlier, check out any of these on Google Scholar:

- Dangerfield J.M., Telford S.R. (1993) Aggregation in the tropical millipede Alloporus uncinatus (Diplopoda: Spirostreptidae). Journal of Zoology 230: 503-511

- Dangerfield J.M., Milner A.E., Matthews R. (1993) Seasonal activity patterns and behaviour of juliform millipedes in south-eastern Botswana. Journal of Tropical Ecology 8: 451-464

- Dangerfield J.M., Milner A.E. (1993) Ingestion and assimilation of leaf litter in some tropical millipedes. Journal of Zoology 229: 683-693

- Dangerfield J.M. (1993) Ingestion of mineral soil/litter mixtures and faecal pellet production in the southern African millipede Alloporus uncinatus (Attems). Pedobiologia 37: 159-166

- Dangerfield, J.M. & Kaunda S.K. (1994) Millipede behaviour in a savanna woodland habitat in south-east Botswana. African Journal of Ecology 32: 337-341

- Dangerfield J.M., Milner A.E. (1996) Millipede faecal pellet production in selected natural and managed habitats of southern Africa: Implications for litter dynamics. Biotropica 28(1): 113-120

The things we do, hey.

About the Authors

Mark is an ecologist, not a green one. He has built, run and sold companies, held academic posts at four universities, advised governments on environmental matters, has a PhD and still plays too much golf. His ecological research ended up in more than 70 peer-reviewed research papers and over 80 technical reports, and he supplied peer-review for more than a dozen international science journals. He lived in Africa for a decade, where he researched termites and millipedes, and once walked into a lion who snarled at him. Much of his career has been in Australia, where he won an Outstanding Teacher award from Macquarie University.

His consultancy practice advises commercial and public sector clients on assessing, monitoring, and evaluating environmental impacts, especially in carbon asset and liability management. His non-fiction books are about awareness of humans' role in nature and, as Paul Sorol, he chronicles and writes fiction for a change.

Link in bio

Chris is an ecologist, a restoration one. A lifelong mindful sceptic, Chris has happily landed as an ecological practitioner in his bush regeneration business. When not passionately managing land, being awarded Landcarer of the Year, trawling the evidence on nutrition, diet and health or carefully advising NGOs and government, he grows plants in his commercial nursery.

MINDFUL SCEPTIC GUIDES

Abatement

Biodiversity loss

Escaping the Malthusian trap

Farmers

Food security

Food, people and soil

Growing enough food

Healthy and sustainable diet

Sustainability

The soil carbon debt

mindfulsceptics.info

MORE FROM MINDFUL SCEPTICS

Uncover the power of curiosity, navigate challenges with critical thinking, and transform your perspective on how humanity can survive itself with more from the Mindful Sceptics

In posts and helpful content guides, help yourself to redefine scepticism as a tool for intelligent doubt, empowering you to question, evaluate and make informed decisions

Unlock the door to science, evidence and insights to understand the human predicament on a finite planet

Sign up now for exclusive access at

mindfulsceptics.info

Made in the USA
Monee, IL
04 November 2026

67380036R00069